2

The House
at the
End of the Road

✸ Smithsonian Books

HARPER

An Imprint of HarperCollins*Publishers*

The House
at the
End of the Road

*The Story of
Three Generations
of an Interracial Family
in the American South*

W. RALPH EUBANKS

Photo on page iii courtesy of Charlie McCullers.
Portrait of Jim Richardson on page 190 courtesy of the author.
Photo on page 192 courtesy of the author.

HarperCollins books may be purchased for educational, business, or sales promotional use. For information, please write: Special Markets Department, HarperCollins Publishers, 10 East 53rd Street, New York, NY 10022.

Designed by Suet Yee Chong

Library of Congress Cataloging-in-Publication Data is available upon request.

ISBN 978-0-06-137573-6

09 10 11 12 13   OV/RRD   10 9 8 7 6 5 4 3 2

FOR COLLEEN,

FOR ALWAYS.

AND

IN MEMORY OF

JAMES MORGAN RICHARDSON

AND

EDNA HOWELL RICHARDSON

Contents

Prologue   ix

PART I: PRESTWICK

1   The World As They Found It   3

2   Jim Richardson   21

3   Edna Howell   33

4   With Eyes Open and Shut   45

5   Out from the Shadows   59

PART II: REACHING ACROSS THE CHASM

6   A Beautiful, Needful Thing   77

7   Parallel Lives, Separate Legacies   91

8   A World Lost   105

PART III: TRANSCENDENCE

9   Transcending Ambiguity   121

10 Moving Beyond the Myth   139

11   The Next Generation   155

12   New Moon Over Alabama   173

Epilogue   191

Selected Reading   197

Acknowledgments   201

Prologue

"THE HARDEST THING I EVER DID WAS TO ASK A WHITE man to marry his daughter," I heard my father say late one night to my mother when they thought I was asleep. Then, there was a pause, as my father took a drag on his ever-present Salem cigarette. "And I'm not sorry that I had to do the asking. It's all been worth it."

The two of them laughed tenderly, which I felt in their voices since I could not see the expression on their faces. It was 1973, and I was an innocent sixteen-year-old, blanketed by naïveté, lying awake in a room filled with boyhood toys and model airplanes. Though I tried to listen in as my parents talked into the night about how they forged ahead with their marriage in spite of different backgrounds, I remember almost nothing of their whiskey- and nicotine-fueled discussion. Rather, I remember how puzzlement filled the air of my room like the smoke from my parents' cigarettes and crept into my brain as I attempted to process that my mother's much-loved father was a white man.

My grandfather died six months before I was born, so I knew

him only through the stories my mother told. I just assumed that since my mother was black, so was my grandfather. His race was never discussed, and in Mississippi in the 1960s it would not have been discussed without severe social consequences for my parents. Now I understood why my grandfather's portrait sat in my parents' closet, emitting a dusky evanescence from behind the closed door.

Yet that evening, overcome with adolescent narcissism, I thought only about how what I had just heard affected me. The impact on my mother of living in a mixed-race household in the Jim Crow South of the 1930s stood invisibly in a dark corner of the room. With a consciousness rooted in black pride and the civil rights movement, I quickly concluded that my grandfather's race had nothing to do with my racial identity. Soon, I was fast asleep, leaving my parents' conversation to drift into the night air.

Almost thirty years later, memories of that evening began to come back. My mother and I were together with my daughter Delaney in her school for a classroom presentation about holiday memories. After the children told their stories about family Thanksgivings, Delaney innocently asked my mother to tell her about the most memorable Thanksgiving from her childhood. Holding tightly to her emotions, my mother struggled to tell Delaney about a special Thanksgiving when her mother's friend Miss Callie came and made a big dinner just for the children and played games with them. It was painfully clear to me that that was all she could say without losing her composure in front of a kindergartner.

When we left the classroom, we stopped on the front steps of the school on the way to the car. "What happened that Thanksgiving, Mom?" I asked quietly, sensing that she was still upset. Then, standing ramrod straight and glancing directly into my eyes, my mother began to tell me a story I had never heard. This time I did not let my thoughts drift away as I did all those years ago. Instead, I listened intently as the memories etched lines of pain on her face.

"When my mother died, for a brief time I felt like I became invisible," she told me. "And the Thanksgiving I was trying to talk about was when my mother died." My mother, Lucille Richardson, was only seven years old at the time: old enough to know what had happened to her mother but young enough to slither through the rooms of her house unnoticed by her father, who was devastated by his loss. My mother recalled that all the mirrors in the house were covered with thin white sheets, perhaps to keep her mother's spirit at rest. That made it easy to sneak around.

As she surreptitiously watched her mother being placed in a casket wearing a powder blue dress with a white lace collar, a dress her mother had sewn with her own hands, Lucille wondered what would happen to her now that her mother was dead. In spite of her ability to hide and eavesdrop, she struggled to decipher what the grown-ups around her were saying about her future. Then, just days after her mother's burial, she found herself sitting on the backseat of a 1936 Ford with her nine-year-old sister, staring through the car's rear window as her white clapboard farmhouse at the end of the road got smaller and smaller in the distance, finally fading into one of the clouds of dust that billowed behind the car. Sent to live with relatives in Mobile, Alabama, separated from the home she loved, her older siblings, and her father, Lucille did not know when she would return. Before she could come back to her house at the end of the road, her father had to figure out what to do with his family. Would they stay in Alabama, or would they move north to pass for white? It was something that was whispered about among the family, something Lucille overheard and finally connected with an incident from a few years before her mother died.

That day, her father, Jim, was injured in a logging accident a few miles from their house. After the accident, a group of men, all of them white, brought him to the house. Confused by all the commotion and her father's cry of pain, Lucille turned to her mother's friend, Miss Callie, who had helped form her most vivid Thanksgiving memory, and asked, "What's wrong with Jim?" To everyone, including his children, her father was known by his first name, without any pretense of formality.

"Good God almighty, little girl, I didn't ask them what was wrong with that white man," Miss Callie replied.

Confused, Lucille turned to Miss Callie and asked, "Is my daddy a white man?" Miss Callie then shouted with disbelief to

Lucille's mother, "Edna! Why didn't you tell this little girl that her daddy was a white man?"

"Because it's her daddy and it doesn't matter, Miss Callie."

But now, just three years later, whether you were black or white did matter for some reason. The family's future seemed to hang in the balance between the black and white worlds they straddled. The town doctor who pronounced her mother dead offered to help the family start over, far away from rural Alabama, as a white family. Even some of her lighter skinned black relatives talked about how the Richardson family could slip into the white world unnoticed. Although Edna was black, all the children's birth certificates said they were white. So, it would have been easy. In the end, Jim Richardson chose not to hide his children's mixed race behind a lie. The family was reunited, and Jim chose to make his children acknowledge who they were, rather than to see themselves the way the rest of the world chose to see them.

To this day, no one in the Richardson family has regretted making what seemed at the time the harder of two choices. This is the story of why they made that choice and how it has reverberated through the family for three generations.

I

PRESTWICK

A place belongs forever to whoever claims it hardest, remembers it most obsessively, wrenches it from itself, shapes it, renders it, loves it so radically that he remakes it in his image.

—JOAN DIDION

The World As They Found It

THE CANOPY OF MOSS HANGING FROM THE OLD LIVE OAK trees gives this place a torrid primeval beauty. As I walk down the winding sand-covered roads, I'm overcome with a peaceful silence that's interrupted only by the soft murmur of the soles of my shoes moving through the sand. There's not a soul in sight. Then I turn a corner and see it: an abandoned church covered in vines of wisteria, the first sign that there was once a real community here. Around the next corner stands a trailer home with abandoned clutter in the yard. After a few more twists and turns there sits a white clapboard house at the end of the road that everyone around here will tell you was the home of the most important family in this now deserted town of Prestwick, Alabama.

Prestwick disappeared without a trace from the Alabama state map more than a generation ago, about the time the nearby whistle-stop hamlet of Carson met the same fate. In the years since their respective post offices and general stores shuttered their doors, almost every sign of these once vibrant communities faded away. As I wander these roads, I stop and ask people what this place was like years ago. First, they hesitate. It's almost as if the little that remains of the past in the present makes it painful to speak of the distant glory of these communities. After a pensive pause, they start to tell me about those majestic trees that look much as they did years ago. Some even speak of them as if they are the only friends from the past they have left.

The older residents are right: the grandeur of the live oak trees is about all that remains of Prestwick. There are people here, but few of them are young and the older ones are slowly dying. What once brought Prestwick to life has disintegrated into the dusty road under my feet and under the headstones of the cemeteries I pass along the way. The railroad tracks near the site of the old depot bear a brown coating of rust from lack of use. And with strong winds from storms off the Gulf of Mexico in recent years, some of those much-prized majestic trees are fading into memory as well. Storms, combined with developments of large, lavish houses for weekend hunting and fishing parties from nearby Mobile will soon erase what is left of a world lost.

In spite of it all, Prestwick, Alabama, clutches tightly and lovingly to its former glory as a largely independent black community. From 1890 to 1910, Prestwick, like the rest of Washington County, grew at a rate double the rest of the state of Alabama. Still the area remained relatively isolated, with an average of thirteen people per square mile. The community's ancient appearance is just one sign of this once-isolated way of life. Another is the white house on the raised cinderblock foundation at the end of the road just past the now-deserted Mount Moriah African Methodist Episcopal (AME) Church. Like the family that once lived there, the house is the survivor of many a storm. Though rocked and battered, it looks much as it has for almost a century.

The street sign that points the way to the house reads "J. Richardson Road." In this isolated rural setting, a green reflective street sign seems somewhat out of place. But without it, there is no way to know that this was once the home of Jim and Edna Richardson, two of Prestwick's best-known, and some might even say notorious, residents. Jim was born in 1887 down the road in what was then called Carson, a mostly white farming enclave. Carson was one stop down the railroad tracks from Prestwick, an independent all-black community, with the exception of the white-run sawmill. But it was to Prestwick that Jim, a white man, chose to take his wife Edna, a black woman, to live in 1914.

Jim and Edna Richardson were my grandparents. That white clapboard house on the cinderblock foundation is a place filled with stories, mysteries, and secrets. The house remains in the family, kept standing by my first cousins Jimmy and Carolyn Jenkins who live there and serve as the custodians of the memories and history created within its walls. At the time of Jim and Edna's marriage in 1914, interracial marriage was only discouraged by Alabama's state constitution. By 1929, when their last child, my mother, was born, it was declared a felony, with penalties set at two to seven years of hard labor.

Sixty years later, when my wife Colleen and I married, there were no laws standing in the way of a black man marrying a white woman. There were also no laws that would have made our children illegitimate, as there were on the books when my mother was born. While my marriage and family have parallels with my grandparents, the differences to me are dramatic. For that reason, I feel that I owe a debt to my grandparents for the life my family has, one unencumbered by strict laws and social rules. To repay that debt, I decided to reconstruct Jim and Edna's lives as best I could, to tell their stories before whatever they left behind has completely faded away. To begin, I had to come back to Alabama, to what remains of Prestwick and its neighboring town of Carson.

Prestwick and Carson, with only three miles between them and separate post offices, train depots, and general stores, established a dividing line between the black and white communities in this part of Washington County, Alabama. Edna was born in 1895 apart from these two communities, in Saint Stephens, Alabama's territorial capital until 1819 and later the county seat of Washington County. Today, however, like Prestwick and Carson, Saint Stephens belongs to the ages, its glory days having faded into memory soon after the Civil War. Over in Saint Stephens, original French and Spanish settlers, blacks, whites, and Native Americans had intermarried since the first settlers arrived at this crossroads of the old Mississippi territory in 1772. Edna's outward appearance revealed this mixture of cultures.

An exotic-looking woman, Edna had olive skin and long black hair that stretched down her back. Legend has it that her mother Adeline was a Creek Indian. It is more likely that she was a mixture of black, white, and Native American, as are most people of mixed race in these parts. Still, Edna thought of herself as black and by all accounts was a beautiful woman who caught the eye of many of the local men, including Jim Richardson.

Since Jim Richardson lived near Carson, I imagine that he must have encountered Edna on one of the land and business transactions he made in Saint Stephens. I can only imagine this because there are no letters or diaries to recount this meeting or document their courtship. But the story of these two people still resounds in the tales told to me by this now dwindling community of people who remember them. Sadly, I can rely only on these stories and what few paper traces they left behind. To tell their story, I have to become Jim and Edna as I listen to people pour out the remaining tales of their life and times. Each time I arrive at their house at the end of the road, one of my cousins greets me and allows me to roam around as I please. In its rooms and open spaces outside, I leave myself behind and start to seek my grandparents out and inhabit their world. The more I let go of myself, the more I find of Jim and Edna.

When the two of them met, Jim was a young man in a hurry to make his fortune by logging, selling bootleg moonshine, and acquiring land in a remote part of southern Alabama that maintained a frontier mindset well into the early twentieth century. Based on census records that include their ages and their children's ages at the time, when the two of them met Edna was just seventeen and Jim was a worldly man of twenty-four. The family agrees that my estimation is about right.

Jim Richardson grew up around black people, since his father had relationships with black people, as well as a child with a black woman, and by some accounts an entire second family. And Jim maintained family ties with the children of that relationship, whom he saw as part of his family. Perhaps he met Edna through this extended family. Or perhaps Edna was attracted by his flamboyant antics, like riding his horse into the general store in Prestwick on a dare. Without a doubt, Jim had a local reputation as a boisterous adventurer who was quick with a gun and had a sharp tongue that could be biting and humorous at the same time. Edna was known for being quick-witted as well, so the two of them made a good match. In spite of his notoriety, and the objections of his family, Jim loved Edna, and she loved him.

They married in 1914 away from the safety of Prestwick, only to return there forever. Whether Jim knew it or not, or whether he would admit it, Edna served as a stabilizing influence on him. She kept him in line, pushed him to maintain his family ties in spite of the tensions created by their marriage, and made sure that he gave her some stability in case anything ever happened to him. They had seven children that he cared for and raised when Edna died just twenty-three years after they married.

Although no pictures of Edna exist, people around Prestwick still speak of her beauty and inner strength. Clearly, Jim was struck by it too, so much so that he left his white world for her. Edna, apparently as shrewd as Jim, made him build her a house at the end of a long dirt road, one that both of them thought would be a safe place to raise a family.

With laws against interracial marriage hanging over them, Jim and Edna somehow persevered. It's clear that they rarely brought up their different races; remarkably their children even had to figure out the racial difference between them on their own. It wasn't until my grandfather was hurt in a logging accident and someone called him a white man in my mother's presence when she was about six years old, that she realized that her parents were different. Consequently, my mother followed the same pattern with me. When I asked why she had not mentioned my grandfather's race before, my mother's reply was simple: "It didn't matter." She was merely echoing what she had heard from her own parents.

But to some people, Jim Richardson's race and his crossing of racial lines did matter; his close alliance with the black community was never an impediment to his work. Instead, he used it to his advantage. Jim Richardson positioned himself in the community as a powerful landowner and political operative, which scared some people around this part of Alabama. He amassed great amounts of land and property, some purchased and some inherited from his father, and wielded his influence with local people who bought both the timber he sold to local sawmills and his smooth-tasting moonshine. In spite of his unconventional, free-spirited life, people respected him. Given his reputation for having a hot temper, he probably instilled fear along with that respect. What scared white people was not that Jim was married to a black woman; they feared that if he was harmed or killed, one of the largest landowners in this part of Alabama would be a black woman. The title of much of his property was held in Edna's name, not Jim's. To whites who may have been contemptuous of Jim Richardson and his race mixing, it was better to leave that economic power in white hands than to create a powerful class of black landowners. Jim Richardson knew this and used it to his advantage.

Intermarriage, though not accepted among whites, was common in what were then tightly cloistered communities with rules and codes of behavior that belied the broader culture. Many local people tolerated intermarriage, but the state of Alabama did not like the racial ambiguity created by the children born from these relationships and consequently created strict legal parameters to define who was black and who was white. Marriages led to the formation of families, and if interracial families were allowed to become fundamental units of society, the state of Alabama would be sanctioning racial equality. To maintain a racial dividing line that would keep blacks in an undeniably subordinate and inferior status in society, inter-marriage had to be made illegal. Consequently, any child born of an interracial union was also deemed black, inferior, and illegitimate.

By the 1920s, when Jim and Edna's children were coming of school age, in most of Alabama, anyone with a black relative was black, even though their legal designation was "mulatto." "We lived as black people, went to black schools, and thought of ourselves as black," my mother and her sister Mary always remind me. Yet in spite of these legal definitions of race, when you looked at the Richardsons you'd ask yourself, were they black, or were they white? Most people simply could never tell.

"He's just a little sunburned," Jim once noted to a white barber in Mobile who remarked that his son Robert was "kind of dark." Though the Richardson children lived as black people, their ambiguous appearance sheltered them from some of the harsh realities of Jim Crow segregation and enabled them to escape some of the social distinctions that were based entirely on race and skin color. Looking at many of the mixed-race people in Washington County, even today, you can see that there were mixed-race families other than the Richardsons, although I am not aware of any who lived openly as the Richardsons did. Quite often, it's hard to tell who is black and who is white. While the Richardsons were sheltered from the racial conventions of Alabama, the strictures of the American South still intruded on their isolated life, beginning with local customs in the school system.

When Jim and Edna married and settled into the world of Washington County, it already had an almost century-old tradition of interracial marriage that led to a large community of mixed-race people who, though a known presence in the county, largely lived apart from blacks and whites nestled in their own world and culture. Unlike the rest of Alabama, Washington County, along with Mobile, maintained three separate school systems: one for whites, one for blacks, and one for "those of racially mixed heritage." Until the 1960s, many of the schools for black and racially mixed children were small three- or four-teacher schools that taught grades one through twelve, although there were usually only two or three children above the seventh grade.

In the early part of the twentieth century, these racially mixed people were known by locals as "Cajans," a name provided by a state senator named L. W. McCrae, who, after seeing the Louisiana Cajuns, decided that the people he knew back home who looked a bit like them ought to be called the same thing, though with a slight variation in the spelling. With the surnames of Reed, Weaver, Rivers, and Byrd, it's clear that the Cajans of southwestern Alabama are in no way related to the Cajuns of Louisiana. Rather than having Acadian roots, these so-called Cajans descended from freed slaves who inter-married with whites and Native Americans. Some of them are the descendents of Daniel Reed, a freed slave who left the island of Santo Domingo to settle in Washington County, where he became enchanted by a light-skinned slave girl named Rose. Daniel worked and bought Rose from her owner. In 1818, as required by territorial law, he obtained her freedom in writing from the General Assembly, and it is recorded in the territorial records:

> Be it enacted by the Legislative Council and House
> of the Alabama Territory in General Assembly
> convened . . . that Daniel Reed, a free man of color,
> be and he is hereby authorized and empowered to
> emancipate, set free and discharge from the bonds
> of slavery his Mulatto slave, Rose . . .

After Daniel Reed married Rose and they had children, he bought them as well, since technically they were the property of Rose's former owner, Young Gaines. A person of color was not born into freedom in Alabama, but had to petition for it. In the Washington County courthouse, you can still read the bill of sale that Daniel Reed got for his son:

> Know all men by these present that I, Young
> Gaines . . . in consideration of the sum of five hun-
> dred and twenty five dollars paid to me by Daniel
> Reed . . . do hereby give, grant, bargain, sell and
> convey unto Daniel Reed his son, George, for the
> sum aforesaid . . .

Later, after purchasing his own son, Reed petitioned the General Assembly of the Alabama territory for the right to emancipate his children and gained their freedom.

Eventually the Reeds intermarried with whites and Choc-taw Indians, many of whom managed to remain in Alabama after their removal from the region in the 1830s. Given their outward appearance, the Reeds were listed in both black and white marriage records of the late nineteenth century. As the families of these unions grew, they sought to be affiliated with the dominant white society rather than the oppressed black society. Darker members of families were given the label "Negro," while lighter ones were labeled white.

But Alabama's preoccupation with racial purity came to bear on how these people of mixed race, who were a growing segment of the population of south Alabama, could identify themselves in the broader culture. In the 1890s, Alabama defined a mulatto as anyone who was five or fewer generations removed from a black ancestor. By 1927 the state legislature defined mulatto as any person "descended from a Negro," no matter how many generations removed. In the rest of Alabama, where these mixed-race people were few and far between, this legal definition kept mixed-race people defined as black. But there were far too many people of mixed race in Washington County for this definition to work. It didn't matter that mulattos and Cajans had no common ethnic ancestry. But the label "Cajan" stuck and worked to keep these people of mixed race in their place, to the point of virtually isolating them from the broader society. As Richard Severo noted in an article titled "The Lost Tribe of Alabama:" "The Cajans are a people without a race and they have the great misfortune to live in a state where you must have a race so you'll know where to go."

By the 1970s, this community threw off the designated name and is now known as the Mobile-Washington (MOWA)

Band of Choctaw Indians. Since then, the group has, so far un-successfully, sought federal status as a Native American tribe. However, the state of Alabama recognizes the community as a Native American tribe rather than as an isolated, ethnically ambiguous group.

Rather than have their children marginalized like the Cajans/MOWAs, the Richardsons chose a different path. They lived as a unique family at the end of a winding sandy road in southwest Alabama in a black community. Jim moved between the black and white worlds, which the broader soci-ety of Washington County tolerated but did not completely accept. At the same time, Edna raised their children in the black world of Prestwick to give them a sound, unam-biguous identity, so that unlike the Cajans they would know where to go. They were members of the local AME church. The seven Richardson children, though technically racially mixed and in spite of the "white" designation on their birth certificates, all attended the all-black Rosenwald School that was built around the turn of the century and provided an education to the thriving black community that lived there. Rosenwald Schools were funded by philanthropist Julius Rosenwald, as a way of supporting chronically poor African American schools in the South. And Jim, rather than having his children set apart like the Cajans, taught them to move fluidly between the black and white worlds. They were not black, white, mulatto, or Cajans. First and foremost, they were Richardsons.

———

WASHINGTON COUNTY, ALABAMA, WAS AND STILL IS ONE OF the most isolated parts of the state, which helped the region maintain a frontier mentality up until World War II. North of the port of Mobile, the Tombigbee River runs though the county and divides it from neighboring Clarke County, another part of the rugged Alabama frontier. While Washington County had a long tradition of wildcat loggers and moonshiners, Clarke County had an all-out war in the 1890s. Led by a group called the "Hell-at-the-Breech gang," the Mitcham Beat War, as it is known, erupted over a dispute about a farmer's debt to a wealthy landowner. The group was made up of backwoodsmen and small-time cotton farmers who set out to even the score with what some call the "county seat governing class." Accounts of the organization's origins vary, but the gang seems to have come about because of cultural estrangement between farmers and Clarke County's middle-class elite. To show their resistance to the county's establishment, the Hell-at-the-Breech gang extracted blood oaths from the men in the community to join its ranks and launched a series of raids in Clarke County, leaving several men dead in their wake.

The violence brought by the Hell-at-the-Breech gang was influenced by a number of factors—politics, economics, class— that extended into Washington County. Alabama historians refer to that time as the "Turbulent '90s," since lawlessness was rampant in this part of south Alabama. The lawlessness was driven by the region's isolation as well as by politics, the economy, and social conditions that were the same in both Clarke and Washington counties. Both places had a culture rooted in white supremacy and middle-class values, with little tolerance for those who lived outside those core values. Consequently, the class tensions that led to the Mitcham Beat War were part of the world that shaped Jim Richardson.

For many years the Tombigbee River served as the region's only passageway to the wider world. Later, the Southern Railroad connected many of the south Alabama communities to each other and to the rest of Alabama, leaving them less insular. It was in exploring this wider world that had opened up, on the rails and horseback, that Jim Richardson met Edna Howell.

When they married, Prestwick was a unique place, one not linked to the violence of the 1890s. The Richardsons quickly became well known in Prestwick, since Jim was the only white person who lived there. While moss-covered live oaks lined the roads around Prestwick, the surrounding woods were full of longleaf pines, making the town a center for logging and giving it a busy sawmill. Until the early 1950s, the Southern Railroad came through Prestwick, carrying passengers twice a day. The school principal even doubled as the postmaster and the operator of the general store and train station. Life was simple enough that everything you needed, you could get in the little town store. If you couldn't get it there, you really didn't need it.

The world of Prestwick as Jim and Edna Richardson found it bore the simplicity of most rural communities, yet at its core life was more complicated. Logging tied Prestwick to a wider world, while its remoteness from that same world helped maintain its innocence. Written records in the community were rare. The oral tradition was and remains of importance in this part of Alabama, a tradition that is both rich and jealously guarded from people outside the community. Still this oral tradition gives a sense of the cultural milieu that shaped Jim and Edna's lives, and it is this tradition that I must rely on to tell their story.

What complicates Jim and Edna's world is the tangled history of race and culture and their respective families' roles in that history. The Richardsons and the Howells, though from worlds separated by race and class, had been on a collision course with each other from the time of their arrival in this remote part of the Alabama frontier. And Jim's family had a history in Washington County that went back before the section of the Mississippi Territory that eventually became Alabama was part of the United States. Yet in the eyes of other members of his family, as well as the broader society of Washington County, Jim's legally black children did not share in the family's deep legacy in the county. Race separated them from what some in the family regard as an aristocratic heritage. Still, Jim taught his children that they were Richardsons first, so that they would always know that this legacy belonged to them as much as it did to him.

When Jim Richardson died in 1956, almost exactly six months before I was born, some of the ties to the Richardson past were lost as well. And I lost a tangible, physical link to my grandparents, two people who have remained distant, unknowable spirits for my entire life. It is as if they have both been locked away as well as hidden in plain sight. As I wander through Washington County, Alabama, on my quest to find Jim and Edna, I also struggle to gather together those fragments of the past and reclaim them. By finding Jim and Edna, I hope to find a piece of my own interracial past, one that I can share with my children to tell them more about these close yet distant ancestors. It doesn't take me long to realize that it is only by reclaiming pieces and fragments of the past that I can find out who Jim and Edna Richardson really were.

The White and colored people of this country [can] be blended into a common nationality, and enjoy together . . . under the same flag, the inestimable blessings of life, liberty, and the pursuit of happiness, as neighborly citizens of a common country. I believe they can.

—FREDERICK DOUGLASS, JUNE 1863

2.

Jim Richardson

A s the train moved down the track from Prest-wick to Chatom, Jim Richardson heard a commotion in a nearby car. "Get out of here, nigger!" a man screamed. In an attempt to avoid a confrontation, the black man who was the object of this wrath tried to ignore the man and his companion, who were both standing over him menacingly. He knew that with one wrong move, these two towering white men would surely beat him to death.

The year was 1913, and Alabama would not enact Jim Crow laws requiring segregated public accommodations until 1928. Still, local custom often required that black people move whenever a white person asked them. But this man was not moving, in spite of the threats from the two men standing over him. He quietly held his ground.

In less than a year, Jim Richardson was to marry Edna Howell, a light-skinned black woman from nearby Saint Stephens. Because of his lifelong relationship with black people, as well as his current courtship of a black woman, Jim Richardson could not sit idly by while these two white men bullied the only black man on the train. So, he did what he thought was the right thing to do: he decided to go to the black man's aid and see if he could resolve the dispute.

But the situation was far worse than the shouting indicated.

The man, already panting from fear, snapped when he saw yet another white man enter the car. More than likely, he thought that Jim Richardson was just another redneck who had come to help his friends finish what they had started. Rather than just lie back and take what he was certain would be an inevitable beating to death, he decided that he was going down fighting.

Jim pushed the two white men to the side, moving swiftly to see if he could help the target of their wrath. But Jim's dominant, take-charge attitude, coupled with his white skin, sent the wrong message. Before Jim could say anything, the black man pulled out a knife and stabbed him in the shoulder.

Soon after, shots rang out. Minutes later, the black man was dead. Jim had killed him. "Not a day goes by that I don't see that man's face," Jim later recounted to the son he called "Smokey," because of the café-au-lait color of his skin, which was darker than that of his other siblings. Smokey tells me that shooting the man haunted Jim so much that he often felt compelled to sleep with the lights on. When I heard the story, it started to creep through my mind as well, one night even penetrating my dreams. I imagined how quickly the events must have unfolded, and the anger coupled with remorse that consumed Jim Richardson at different times. Subsequently, I asked others what they knew about this incident. Interestingly enough, everyone told me the same sequence of events.

Rather than just rely on stories passed from one generation to the next, I thought I would try to find a paper trail that surely would exist for this fatal shooting. I looked through newspapers and documents from 1910 to 1920, since I had no reliable means of knowing exactly when this event took place. Nothing turned up in the local newspaper, *The Washington County News*. In the court records there were numerous Richardsons who turned up for charges ranging from trespassing to assault, including one incident involving Jim's cousin A. G. Richardson, Jr., who was accused of "presenting firearms at another." But only one murder trial appeared in the records, and it did not involve Jim or any other Richardson. Judging from the court records of the time, it seemed that a black man's life was meaningless. In the one and only murder trial I found, black witnesses were used to seal the prosecution's case against two white men involved in murdering a local white farmhand. How could there have been no consequences for the death of the black man Jim Richardson killed?

I found it difficult to believe that Jim just walked away from the murder, leading me to ask more questions as I went from house to house in and around Prestwick. Eventually, I discovered that Jim was indeed arrested and jailed for the murder, but his family lobbied for and gained his release. When I went to search for arrest records rather than accounts of a trial, I still found nothing. "They probably just threw those records out when they released him," one source told me. "Washington County was still the frontier back in those days, and that's just what they did." Consequently, Jim's only punishment was to carry the memory of what he did with him for the rest of his life.

Several of the people who separately told me about this shooting believed Jim constantly tried to make up for it. He was a major donor to Mount Moriah AME church, though an irregular churchgoer. Black men were his most trusted work companions and held major responsibilities on his logging crew, which pulled mighty pines and virgin cypress from what was left of the Alabama timber frontier. Jim's temper, antics, and run-ins with the law are legendary in Washington County, yet few written traces of them exist. Jim artfully swept his tracks clean, leaving few signs of his presence. Although he clearly preferred to keep his past shrouded in mystery, the written record reveals much about the family that shaped him. What I discovered was that it was the prominence and standing of his family that in many ways set the path for the life he chose to lead.

While Jim's adult life was spent straddling the black and white worlds, his family's background bears the mark of the white landed gentry of Alabama's early frontier. The Richardsons' roots in Washington County stretched back to the original British loyalists who settled there in 1772 and brought their prim Tory sensibilities to this wild frontier settlement. His great-great-grandfather, Thomas Bassett, held a land grant from King George III. That grant of a 1,050-acre tract read

> . . . [we] do give and grant the described lands together with all woods, underwoods, timber and timber trees, fishing waters, water courses, profits, commodities, hereditaments and appurtenances thereunto belonging, also with privilege of hunting, hawking, and fowling upon same, reserving to us, however, all mines of gold and silver.

There were no mines of gold and silver on this tract of land, but the family kept it nonetheless. Legend has it that Thomas Bassett was killed by a band of Native Americans around 1781, near his cherished property. In spite of the tragedy, the family held on to the land, which today holds a historic marker that calls the spot "Bassett's Creek." After the defeat of the British, Thomas Bassett's sons, Nathaniel and Thomas, became Spanish subjects and obtained Spanish grants to the same tracts of land that they held as British subjects. Later, when the territory came into the United States in 1804, Thomas Bassett petitioned for rights to the property. To prove his case he submitted notarized depositions, translation of Spanish titles to the land and other documents. His claim was approved by the United States Commissioners, who noted that each claim "was supported agreeably to the requirements of the law." By the time Thomas Bassett's daughter Lucinda married John Richardson, a cotton gin manufacturer from Georgia who became a slave-holding planter in Alabama, the family had been major landowners in the county for three generations.

Jim's father, James Morgan Richardson, was the grandson of John Richardson. James's father was John Augustus Richardson, who, like his father, was a wealthy landowner who also owned slaves. True to the family's pattern, John Augustus Richardson's life was not without its scandals. In addition to the five children he had with his wife, Sarah, he had six others with a mistress who was known in the family as "the widow Charlotte." The widow Charlotte caused a great deal of acrimony in the family when John Richardson died and he left part of his estate to her. Two of his six children with the widow Charlotte are buried apart from the other Richardsons in the Pine Grove Cemetery in Leroy, Alabama. In part, because of the family scandal, they could not be buried in the family cemetery in the Rutan community near Chatom, nor could they be buried near the descendents of James Morgan Richardson in Pine Grove, even though one of the sons, Wade Hampton, was a proud soldier of the Confederacy who was wounded in battle. In Pine Grove, his grave sits apart from those of the other members of the Richardson family. The acrimony in the family still stares you in the face each time you enter the cemetery.

When Jim Richardson was born, and named for his father, James Morgan Richardson, the Richardsons viewed owning property, and lots of it, as a birthright. Following his ancestor's example, Jim's family taught him that the land, timber, and fishing waters were as valuable as mines of gold and silver. There was also a pattern of tension and squabbling in the family, though no family likes to keep acrimonious stories alive, particularly when they reveal a family weakness. Nonetheless, I kept stumbling across stories of family disputes over money and property, even though Jim Richardson's contemporaries have long since been buried under the sandy south Alabama soil.

One hot sticky June afternoon I drove around Washington County looking for stories about Jim Richardson and his family. The shooting on the Southern Railroad always came up, as did a few stories of Jim's marriage to Edna. Both distant and close relatives talked of a Jim's mother's disinheriting him because of his relationship with Edna Howell. Soon, I found myself in the Washington County Courthouse in Chatom, digging around for records of this disinheritance.

It didn't take me long to find something.

I had been told that Jim and Edna married in 1914, and the date of the murder on the Southern Railroad was 1913. In the miscellaneous legal records of 1913, I found an entry that said "Kind of Instrument: Release. Date: April 3, 1913." That was the day that Jim's mother, Henrietta "Etta" Powell Richardson, released, that is ended, all regular payments from his father's estate to Jim. The wording of the document was short and sweet: it simply stated that all regular payments from the estate to James Morgan Richardson would cease.

By this point, Etta Powell had remarried and kept the marriage in the family: she married William Henry Richardson,

Jim's father's brother. Following the family's emotional history, there was no question that this marriage created some tension. As I followed the paper trail back a few years I discovered that a year before his death in 1902, the senior Jim Richardson repurchased part of the original land deeded to his ancestor Thomas Bassett, which was rich in timber. Now, this land, the same land Thomas Bassett once leased to Etta Powell's ancestor William Powell back in 1787, was part of the estate and the children of Jim Richardson were entitled to a share of it. The land became a pawn in Etta's attempt to control her son's life. And this was not the first time Etta Powell Richardson sought to assert her dominion over her son.

Around 1910, a few years before his mother disinherited him, Jim Richardson had decided to marry the young daughter of a tenant farmer. Poor whites were thought to be of an inferior, boorish lot by upper-class land-owning whites such as the Richardsons, a sensibility Etta Powell Richardson embraced even though she herself came from a poor white family. Class being almost as divisive an issue as race, Etta was furious. She had married up, and she would not allow her son to descend to the depths from which she came. However, Etta did not try to cut her son off financially. Instead, she paid to have the young woman's family effectively banished from Washington County. As additional punishment, Etta beat her son with a horse whip, loudly shrieking her disdain and disapproval of his choices. Jim, however, steered clear of his mother's clutches and took control of his life when he married Edna Howell. And in the years that followed their marriage, Jim began to repurchase his inheritance by buying land from his family's holdings.

Though land records exist, there are no papers indicating a legal marriage in the state of Alabama. Some people say that Jim and Edna went away to Mississippi to be married in secret, since no one would know them or Edna's background there; others say they had their marriage blessed in the Mount Moriah AME Church in Prestwick. I believe they went away, since Jim knew that he had to avoid his mother's temper and her facility with a horse whip. And at the time Mississippi was a haven for quick marriages, since the state required neither documentation nor a waiting period. Whichever case is the actual one, in the eyes of the state of Alabama, theirs still was not a legal marriage. Subsequently, all of their legal land deeds refer to "James Morgan Richardson, an unmarried man" or "Edna Howell, an unmarried woman."

In the eyes of Alabama law, Jim and Edna were not married. But in their hearts, they were united.

TO AN UNTRAINED EYE READING THE LOCALLY SANCTIONED history, it would seem that only one branch of the Richardson family lived in Washington County. In this history, all those Richardsons are portrayed as noble early settlers, enterprising business owners, and publicly proclaimed pillars of the community. Needless to say, the more scandalous, racially mixed Richardsons are not featured in the pages of local history and for obvious reasons. For historical purposes, the lineage of the Richardson family, as well as most white Southern families, carries a "whites only" sign. Even though Jim and Edna's children are direct descendents of one of the first families of Alabama, for purposes of sanctioned history, it's as if their line of the Richardson family never existed.

Yet, to most white people in Washington County, and some blacks, Jim and Edna Richardson and their children were neither black nor white; they were just known as "those Richardsons." The couple refused the local custom of designating their children as racially mixed. And in spite of the children's racial designation as white on their birth certificates, they also refused to identify themselves as white in spite of their outward appearance. Instead, the family existed as an entity unto themselves, living as a black family that moved between the black and white worlds, rather than sealing themselves into the boxes that local people wanted to fit them in. Perhaps that stand played a role in denying them a place in local history. Although "those Richardsons" may not be present in the annals of local history, the active oral tradition of the American South has kept their life and times alive among the people.

It is this oral tradition that I leaned on to find out more about Edna Howell. Jim's family of white landowners had some written history, given that they were a literate, somewhat aristocratic family. The Richardsons' written record tries to airbrush out some details, but now that so many family members are toward the end of their lives, it is not difficult to uncover what some once wanted to keep concealed. On the other hand, Edna's ancestors, former slaves and Native Americans, left behind few written records. It was through conversations with local people that I learned that Edna's family was as much a part of Washington County's early history as the Richardsons. What happened when Jim and Edna married was the inevitable linking of two historically significant families of Washington County: one white and one black.

A thing of beauty is a joy forever: Its loveliness increases; it will never pass into nothingness.

—JOHN KEATS

3.

Edna Howell

STORIES FLOW IN ABUNDANCE WHEN JIM RICHARDSON'S name is mentioned around Prestwick. But when you bring up Edna Howell, there is often silence, or hesitation. After a few Jim Richardson anecdotes, people will tell you that Jim Richardson had a wife, that she was black, and that she was exotically beautiful. Yet only a select few know her name, who she was, or where she came from. It's almost as if she lived her life so quietly that only faint whispers of her existence can echo from the past. Perhaps she wanted it that way.

To learn about Edna Howell, and not just about the man she married, I knew that I had to dig deep and hard. The narrow bends of the sandy roads of south Alabama bore few signs of her presence. The only way to piece together Edna Howell's life was to begin at the end: in the cemetery of Mount Stoney AME Church. On her tombstone, you see that she died at the age of forty-two. So, Edna's faint imprint on Prestwick is a result of the misfortune of dying young. Because she was so young when she died, my mother remembers little about her mother and my grandmother. And now those who held the strongest memories and knew her best are also dead: her sisters and eldest children as well as her friends and confidants.

Even after hearing a few stories about Edna's life, the first piece of information about her I found and that gave a sense of who she was beyond various oblique physical descriptions was engraved on her headstone. Bearing the weathering of almost seventy years, it states, "She was the sunshine of our home." Stumbling across the grave, I wistfully read its inscription over and over, asking myself what this told me about the woman whose remains lay underneath my feet. Was this inscription a hollow platitude or was there an element of truth in these words etched in stone? Then I thought of the details my mother shared with me about the day Edna died, November 17, 1937.

It was almost Thanksgiving, and Edna was ill with what was called eight-day pneumonia, which today would be fully treatable with antibiotics. Each day, my mother's brother, Edward, drove across a toll bridge to nearby Clarke County to bring their doctor, Jim Bedsole, to tend to her, although at the time there was little that could be done medically for her condition. The toll was a quarter, a sizeable sum in the Great Depression, and Edna always said she would never live to see the day when you would not have to pay a toll to cross the Tombigbee River. On the eighth day of her illness, Edward came back with the doctor and brought his mother some news: the toll bridge was now free. She died soon after. To this day my mother says it was one of the saddest days of her life. The regular rhythm of the Richardson household disappeared with Edna's death, with Jim struggling for years to find ways to bring it back.

Almost seventy years later, telling that story still brings a pained expression to my mother's face, revealing the confused and grief-stricken little girl she once was. I knew that Edna's seven-year-old daughter, my mother, felt a light fade from her life that day. After standing in the silence of the graveyard, without the sound of a single car passing by, I knew the inscription was entirely fitting and appropriate.

To make Edna's light shine again, and even brighter, I left the cemetery in search of clues that would tell me more about her family, her background, and the events that shaped her life. Somewhere there had to be signs of what led her to marry Jim Richardson and to embark on a life straddling the racial divide. Her life was bound largely by domestic duties, like many women of her day, which enabled her memory to fade and to leave no mark on the wider world. Yet somehow I knew Edna Howell was a strong woman; to live with Jim Richardson, she had to be.

I began to delve deeper into Edna's background by exploring her hometown of Saint Stephens. Without letters or archival resources in the Saint Stephen's Historical Commission to guide me, I talked with anyone who would tell me about local history, particularly of the town's African American population. The head of the historical commission told me that the keeper of local black history was a man by the name of L. J. Williams and that perhaps he could help me. Although nearly eighty years old at the time, Mr. Williams looked twenty years younger and, after a few probing questions, he easily identified which branch of the Howell family that I was trying to find. As we talked about Edna Howell, he explained to me how he was related to her father (Gilbert Howell was his father's cousin) and that we had an ancestor in common. Intrigued by this newfound relation, I asked him to tell me more.

Edna Howell's ancestors were African American and Native American, who, in all likelihood, were part of the same tribe that killed Thomas Bassett. Some of the family may even have been owned by Bassett and his descendents, including the Richardsons. The Howells were an amalgam of Creek Indian, African, and European blood and had once been held as slaves. But one member of the family, to the pride of the rest, gained her freedom.

On a high hill behind the main business district of Saint Stephens in the early nineteenth century lived a woman by the name of Mahala Martin, known to everyone in town as Aunt Hagar, who is identified in land records as a "free woman of color." Her owner Solomon Anderson deeded the land she occupied to her along with her freedom in 1820, which caused quite a stir, since a black person, much less a woman, had never owned property in Alabama. When the town fathers of Saint Stephens tried to take Hagar's land, she went to court, defended herself, and won the case, making her one of the first free women of color to own property in Alabama.

Hagar's story of her defense made the front page of the local newspaper, *The Halcyon & Tombeckbe,* nestled between what now appear to be trivial announcements of local and national news. But the story of Hagar's bravery and determination did not make it to the next generation of Howells, although it is preserved in the written histories of Washington County. For the Howells, slavery's sting in any form was something to be forgotten rather than remembered. The history of Washington County prominently and proudly acknowledges that one free black woman lived in Saint Stephens. One oral history read at the centennial celebration of Saint Stephens in 1899 went a little further:

> Well, I would not be true to myself and perhaps not to the subject in hand were I to omit the mention of one individual, whose condition was unique for those days—a free Negro—Aunt Hagar. She lived on a high hill back of the business streets. She was a law abiding citizen, cultivated her own little garden and patches; had her cow and chickens, and enjoyed the full confidence of whites. How we, the little ones, did love to visit her, with or without permission, as the case might be! She always gave us a kind reception and oftimes a treat. Looking back through all the intervening years, I can truthfully say that a slice of her corn hoe-cake split open and spread with butter of her own make tasted sweeter to me then, than the nicest cake does now. How much of the happiness of our lives depends upon

association. She passed away before or during the Civil War. Peace to the memory of old St Stephens' humblest citizen.

Today, Aunt Hagar is played by a reenactor on Old Saint Stephens Days every October in recognition of her historical significance. The Richardsons may be among the first settlers of Washington County, but Edna Howell's ancestor also left her mark in the heart of the community, much as she did on the members of her family.

After leaving Saint Stephens, I began to search for relatives of Edna's parents. Her father, Gilbert Howell, had relatives who still lived in Saint Stephens, yet I could find only a few of them and no one who remembered Edna. Edna's mother, Adeline King, had relatives scattered around the country that I also had to track down. One of them, Edna's cousin, Tom Pruitt, described her just as everyone else I met did: "She was a red-skinned lady, kind of a light brown, very attractive. She had long dark hair. She stayed home and raised the kids, was strictly a housewife. There was nothing else she could be." But something else in that conversation made me realize that Edna was more than a housewife. Tom Pruitt remarked that although Edna was black, she had a house that was the envy of some white people and that she also had as much household help at her disposal as any white woman would have had. That comment led me to realize that Edna had real power in her re-lationships, counter to the prevailing notion that a black woman married to a white man always played a subservient role.

How did a black woman in Alabama in the early part of the twentieth century command what, at the time, was considered a grand farmhouse and domestic help to keep it running? Either Jim Richardson gave this help to her to maintain his status or Edna insisted on it. Though I'll never know, I imagine the status of the family's house with domestic help derived from a combination of Jim's ego and Edna's demands, demands that she be treated like the wife of a man of Jim's stature in the community.

But did he love her? That was something I had to know, to find real evidence, since some whites in Washington County remain incredulous that an interracial couple would live openly as a family. Moreover, many can't let go of the prevailing no-tion that a black woman always plays a secondary role in an interracial relationship, with an unseen white wife looming over her. "He must have married her just to spite his family," some people would say as I recounted my grandparents' story. Eventually I found evidence of the nature of Jim and Edna's relationship not in a letter but in one story of their marriage that I confirmed with others around Prestwick. The story was passed down from Virginia Powell Bryan Sullivan, a cousin of my great-grandmother Etta Powell Richardson, to a local gentleman whom I encountered on one of my Alabama treks by the name of Joe Franklin.

According to Franklin, "Miss Virginia," as she was known, related that as a young man, Jim courted several young women, none of whom met with the approval of his mother. Then he started seeing Edna Howell and became enchanted with her, later telling his mother that he was determined to marry her. Miss Virginia said, "He told his mother he was going to marry this black girl and he did. They raised a family. She died as a young woman, and he continued raising the family. Now this wasn't some affair between a white man and a black woman like you read about in some novel. God knows he loved her."

People I talked with characterized Edna as independent, stubborn, and spirited. "Shrewd," the one word that I would use, never came up. As I glance at the land records of the house and forty acres deeded to Edna by Jim, I imagine the house was built as a condition of Jim and Edna's marriage as much as it was a sign of love. Later I learned that the house came to be built under more complicated circumstances.

Edna's responsibilities included not only taking care of the home and the children, but also keeping Jim's logging crews fed. The first house they lived in together, which was near the house they eventually built, was small and cramped, much too small for a growing family and groups of loggers coming in and out. Once, feeling she was being taken for granted, Edna left Jim and took the children to live with her sister Jannie in Mobile. For years, my aunt kept the letters Jim sent Edna there but, to my great sorrow, finally discarded them in the last few years, thinking they were of little consequence. In one letter from around

1923, Jim begged Edna to come back and promised to build her a house as well as to get her the help she insisted she needed. The logging business was important to Jim, but not as important as Edna. Edna insisted on being Jim's first priority, and after this spat that is how Jim treated her. In return, she focused on raising a family and being the light of their home.

Jim probably met Edna through his social network, which largely consisted of black men who worked in his logging business. Local legend has it that Jim was a fixture at the Friday and Saturday night social functions these men conducted with some regularity. In all likelihood, one of them introduced Jim to Edna. But Edna would not allow herself to be used as a paramour, as many black women were at the time. She insisted that Jim court her and treat her as he would have treated any white woman in his circle. Edna would accept nothing less.

I believe the life and relationship of Edna's parents shaped how she chose to forge her relationship with Jim Richardson. Edna's family scattered when Edna's mother, Adeline King, discovered that her husband, Gilbert Howell, had another family in a nearby town. After their separation, Edna had few ties with the Howells. Adeline, with the help of her family, the Kings, chose to raise her three girls, Jannie, Mollie, and Edna, on her own. Adeline felt wounded and betrayed by Gilbert Howell and consequently taught her daughters to be mindful of men and their promises.

Edna carried that lesson into her relationship with Jim Richardson. Gilbert Howell, though of mixed-race ancestry, lived his life as a black man. He left Adeline with very little. So, Edna knew that given the racial politics of the time a white man would think nothing of leaving her penniless if the relationship did not work, much as her father had left her mother. With the Howells, there was at least some support from the family; there would have been no support at all from the Richardsons. To avoid her mother's fate, Edna made sure that she got a house, property, and household help.

One of the women who helped Edna keep house was her closest friend, Miss Callie. She tried to maintain order in the Richardson household in spite of the chaos and sadness that came after Edna's death. Miss Callie's parents had been slaves and were connected with Edna's vast but scattered family from Saint Stephens, thus allowing Edna to maintain a link to the world that shaped her. Jim's sister, Claudia, sought to impart to Edna how to maintain a household, as it would have been done in the white world, and became Edna's primary link to Jim's family. In turn, Edna served as Jim's tie to the all-black town of Prestwick, teaching him as much as she could about navigating in her world as Jim's sister Claudia taught her about living in his.

While Edna moved between the black and white divide of Washington County, her sisters migrated to nearby Mobile, where they began to move into Creole circles that wavered fluidly between the black and white worlds. In the early 1900s, 44 percent of Mobile's population was African American, with a sizeable number of that group being light skinned or "Creoles of color." Although Mobile was as segregated as any city in the South, there were so many light-skinned blacks of this Creole class and intermarriage of Creoles and whites that there was a blurring of the color line that made segregation difficult to enforce. Some even say that Edna's sister Jannie would sometimes pass for white, which would not have been at all unusual at the time.

But Edna and Jim did not try to fade into the white world. Their family moved between the two worlds, while creating a world of their own: one in which their children would see the racial divide clearly and seek to bridge it at the same time. By all accounts, it was Edna, not Jim, who was the mastermind behind this social strategy. While Jim is legendary for his flamboyant personality, Edna brought to bear a level-headed calm and structure that ordered the life of the family. Jim lived twenty years longer than Edna, yet it was the structure Edna established that allowed the family to live on after her. That is her legacy.

Memory believes before knowing remembers. Believes longer than recollects, longer than knowing even wonders.

—WILLIAM FAULKNER, *LIGHT IN AUGUST*

4

With Eyes Open and Shut

"WHITE FOLKS KNOW HOW TO TREAT EACH OTHER," AN older white man proclaimed in a friendly yet authoritative tone during one stop in my travels around Washington County. Secrecy and discretion remain central to the local oral tradition, so this man spoke with me on the condition that I would not identify him. "They also know how to treat black folks, and black folks know exactly how they are going to do them," meaning that blacks know exactly how whites will mistreat them. "But God help someone who falls outside of black and white or even claims to," he continued. Though not surprised by this statement, given that he was a white Southern man well into his eighties, nonetheless I was caught off guard. Still, I knew the man's words revealed some truth as well as the complexity and simplicity of how race was viewed in Edna and Jim's world and even today. In Alabama, the racial die was molded, formed, and fired into black and white. There was no room for those who resided outside the forms that had been cast.

As I probed deeper during our conversation, I quietly pondered the man's message, blurted out in his folksy yet jarring statement. Soon he revealed that, as far as he was concerned, people who had their grounding in two worlds or cultures were suspect in the South. "Always were, always will be." After some back and forth, he noted that he was only trying to reveal to me the discomfort some whites, both past and present, felt about people of mixed race. That way of thinking prevailed in the world that shaped him and influenced the mindset of many in his community in neighboring Clarke County, just across the Tombigbee River from where the Richardson children grew up. The same mindset lingers today throughout Alabama whenever the topic of interracial relationships and marriage enters a conversation. At the core of this way of thinking was the belief that people of mixed race were suspect because they knew the cultural underpinnings of both the dominant culture and the minority culture and could move easily between the two social realms. And that scared some people.

This conversation reverberated through my consciousness for days, eventually reminding me of the same fear and ambivalence about racial identity William Faulkner captured in his novel *Light in August* through the mixed-race character of Joe Christmas. By virtue of both his status as an orphan and his ambiguous, mixed-race heritage, Faulkner describes Joe Christmas as "his own battleground." That characterization derived not just from Faulkner's imagination, but mirrored the culture of the South that Faulkner knew and that shaped the upbringing of the Richardson children. Rejected as white, as black, and as a human being, Joe Christmas lives on the margins without social status. In many ways, he represents the way the American South viewed a person of mixed race: lonely and alone. Though Faulkner's descriptions are more eloquent than those of my Washington County companion, nonetheless they echo the same sentiments: "In the wide, empty, shadow-brooded street he looked like a phantom, a spirit, strayed out of its own world, and lost."

Joe Christmas is a fictional character, yet his inner conflict captures how the South tortured people of mixed race by marginalizing them and making few if any adjustments in society for their acceptance. Jim and Edna Richardson did not have to read Faulkner to know this. They lived this life firsthand in a world where your place was determined by race. Of course, Jim understood the South's racial dynamic in ways that differed from Edna. Edna knew what it was like to be cast to the margins of society. Conversely, Jim knew what it was like to be part of the group that dominated the culture and made people of mixed race into their own battlegrounds, bending and breaking them. He'd experienced this in his own family.

Jim's father had two families: a black family and a white family. The two wives did not acknowledge each other, yet for some reason their children did. Jim was especially close to his half sister, Lola. I believe that Lola made Jim aware of the conflict and hurt his father inflicted on her family through a relationship that was not publicly acknowledged and was consigned to the shadows. More than anything, it was his father's dual existence, where he maintained a safe distance from his black wife and mixed-race children, that made Jim choose to live openly with Edna, even though it meant cutting himself off from part of his family. The dominant culture of the South told Jim to follow his father's path, since it was socially acceptable to have a black mistress who was treated as an invisible entity. Instead, Jim chose to defy both the culture that shaped him and the example his father had set for him.

Neither Jim nor Edna was willing to compromise or fall victim to the dominant forces around them. They chose to empower their family through their example of courage in the face of a culture that made many cower away in fear when they violated its strict tenets. Their brilliance as a couple lay in their social cunning. First, they chose to raise their family in a spot that would physically separate them from white society, while at the same time introducing their children to the white world of those relatives who were not hostile. This helped teach their children to behave in a way that recognized racial lines, while at the same time transcending them. Instead of falling victim to fear and internal turmoil, Jim and Edna taught their children that both recognizing and ignoring race would keep them from being marginalized in either of the worlds in which they chose to travel. Also, they were taught that society would not make adjustments for them. For the Richardson family to endure and prevail, Jim and Edna had to make those adjustments them- selves, assert their right to define themselves, and teach their children to do the same. Consequently, Edna made their home a safe haven by her keen focus on domestic life. In turn, every day Jim Richardson openly fought the battles with men who sought to keep his family on the margins of society. He took them into his confidence, earning the respect of some and the fear of many. To keep his family safe, he chose his battles carefully and wisely. And that was how the family, in spite of forces to the contrary, prevailed.

CULTURAL OR RACIAL IDENTITY IS SHAPED BOTH BY THE vagaries of personal experience and circumstances. In every family, there is a constant battle between the influences of home and the larger community.

As my conversations with local people reveal, Prestwick was a place that had a major influence on how you viewed the rest of the world, sometimes as much or more than home life. Adding all the influences together and combining them with a life of moving between two cultures, as the Richardsons did, might seem to lead to a chameleon-like sense of self. Instead, Prestwick served as an anchor for the family's core identity, in spite of white privilege that they may have been exposed to from their white Richardson relatives. Even a third racial group, the so-called Cajans, one labeled by local people as existing outside the traditional boundaries of race, did not confuse the Richardson racial identity. Their core identity, both in the community and at home, was strong enough to withstand these conflicting forces.

"Our shared humanity gets severely challenged when the manifold divisions in the world are unified into one allegedly dominant system of classification," philosopher and economist Amartya Sen noted in his book *Identity and Violence: The Illusion of Destiny*. Further, "The uniquely partitioned world is much more divisive than the plural and diverse categories that shape the world we live in." South Alabama is one of the most uniquely partitioned places in the world, making it a challenge to a family like the Richardsons, who in many ways were trying to develop a shared sense of humanity between the black and white worlds. The unique circumstances of the Richardson family blurred local definitions of race and identity. Sometimes it is clear how the family carefully and perilously guided itself through this racial morass, and other times it is not so clear at all.

Outwardly, it appeared that the Richardsons zigzagged between two racial worlds or at the very least straddled them. Documents from the 1920 census partly reveal this, since in those records Jim Richardson was recorded as the head of household for a white family, yet no children's names are listed in the census. Ten years later, again the census taker marked Jim Richardson as the white head of the household. But this time the census taker listed the names of his seven children as white, and marked his wife Edna as the cook. *"What else could she be?"* I imagine the census taker thinking. This reddish-brown colored woman taking care of all these fair-skinned children who look white could only be a servant. Or perhaps the designation of Edna as the cook was just another ruse for Jim Richardson to avoid scrutiny by the state of Alabama.

Whatever the case may be, the census document provides one view of the tightrope walk the family faced in a world tightly demarcated by race. The outside world examined the Richardsons through the traditional racial boundaries of the South, yet the family chose to live their life as if those racial boundaries did not exist. "Jim wasn't a white man; to me he was just a person," my mother once recounted with resounding passion in her voice. In fact, she was almost indignant that Jim would be defined by his race. "We didn't think entirely in terms of race in our household, but I imagine that is what other people did," my mother once told my daughter. That was just the way it was: the family lived their home life with a marginal consciousness of race, even though census takers and other people in the community looked at the family completely through a racial lens.

Even the threat of violence from the Ku Klux Klan didn't stand in the family's way. When the Klan came to burn a cross at the Richardson house, they were blasted back down the road by Jim's shotgun. The story goes that he shot to kill even though some of the men hidden by the white robes were members of his own family, as closely related as first cousins. Openly or in secret, people could threaten, whisper, and gossip, but their opinions were effectively ignored. In spite of it all, the Richardsons maintained an unassailable integrity about who they were as a family, and it was their lack of artifice that helped them maintain some sense of normalcy in a world dominated by race and racial boundaries.

Perhaps this openness combined with skepticism of the broader world helped Jim and Edna Richardson bring up their children with an undying allegiance to family and to their core racial identity. It wouldn't be extreme to say that they lived as citizens of the Richardson family rather than individuals bonded to Prestwick and the community around them. They didn't deny the existence of the negative and positive forces around them or even try to isolate themselves. Jim Richardson's personal and political influence in Washington County simply made it possible for them to dodge the rules and conventions of the time. Plus, Jim used his race and political influence, combined with his reputation as a hothead, to protect his family. "People were just scared of him," is the sentiment I hear echoed over and over again. The fear he generated, whether real or imagined, made people leave the family alone, freeing them from the harassment and persecution other interracial families faced. It also allowed Jim to move fluidly between two worlds, one populated by the likes of white judges, politicians, and powerbrokers and the other one filled with the black and powerless.

It wasn't just Jim Richardson's cunning that made the life he shaped for his family possible. Since its days as part of the Mississippi Territory, Alabama culture prided itself on personal and local control and interpretation of laws, with a pervasive localism that guided communities throughout the state. From north to south, from its borders with Mississippi, Georgia, Tennessee, and Florida, Alabama consisted of towns that existed as entities unto themselves, forming what some call "island communities." "Alabama government works from the bottom up rather than from the top down," retired probate Judge Tom Taylor recounted to me one afternoon as he recollected the glory days of Washington County. The communities in Washington County and around the state simply saw their way of life as superior and unique and believed that any external influences would compromise that uniqueness. While other Southern states attempted to form more monolithic cultures, Alabamians took pride in their individual communities, many of which had unique peculiarities. In spite of changes to the state constitution in 1901 that limited local government control and enforcement of laws, Alabama continued to function as a series of communities. Prestwick was one of those island communities, and Jim Richardson's local influence enabled his family to lead a life outside of the boundaries set by prevailing state laws and customs.

The Richardsons pushed not only the racial boundaries but also the economic boundaries. Jim Richardson's logging business, while successful, wasn't completely satisfying to him. The real money in logging was made by the sawmill operators, not by the loggers. And there wasn't much money to be made in farming the land he already owned. Having been cut off from some of his family's land holdings, he held ambitions of amassing enough money to buy his own set of land holdings that would rival those of his other family members. The real opportunity Jim saw to make his fortune was not in logging but in selling bootleg liquor.

Alabama became a dry state in 1915, well before the enactment of national prohibition in 1920. With the repeal of prohibition in 1933, the state left the question of legal liquor sales up to county governments. Washington County remained a dry county and still is to this day. It was illegal for liquor to cross county lines, so the very laws that were to stop the sale and presence of liquor in Washington County made the business of bootleg liquor both a way of and a fact of life. Jim Richardson saw this as an economic opportunity and grabbed it.

Bootlegging was thought of by locals as an honest and profitable profession as long as you didn't get caught. And given the local interpretation of laws, people were rarely caught. The local sheriff simply chose to ignore bootlegging, as did the local clergy, who only wanted to keep liquor from being sold legally. And the state authorities didn't interfere with bootleggers unless the local sheriff asked them. The people to dodge were the federal authorities, who tried and failed to shut down local stills largely because their owners were always one step ahead of them.

In Washington County, whiskey was one means of generating influence and power. "Back then, you didn't give money to a political campaign, but you did give liquor," an elderly Washington County resident recalled for me one afternoon. To understand this facet of Jim's life, I began to delve into the culture of bootlegging, a world that has since faded away, to be replaced by marijuana. No one talked about bootlegging in its heyday. It wasn't a topic for polite conversation, just as people today quietly acknowledge marijuana sales on the moss-covered roads of Washington County but don't talk about them. Even today it's spotty going to get people to discuss bootlegging, as if the ghosts of bootleggers and revenuers past might rise up and seek their revenge if too much is revealed.

Whether tight lipped or confessional in their tone, everyone I talked with agreed that, without question, Jim Richardson was the biggest bootlegger in Washington County. No one exceeded his reach or influence in the county. They all firmly believed that in Washington County, the politics of bootlegging trumped the politics of race, providing Jim with an added advantage. He sold liquor to the politically powerful and influential in the county, which gave him a level of influence reserved for members of the same class. The financial rewards of bootlegging exceeded his income from logging and farming, his other businesses; the social and political rewards were merely an added bonus. Jim's openness about his life led the political class in Washington County to place trust in him, tell him their secrets, and protect his open secret of bootlegging. Conversely, Jim behaved as if he were just an average, run-of-the-mill type in spite of the complexity of his life. What people often did not recognize was that below that veneer of ordinariness lay a real political operator.

From the time they arrived in Washington County, the

Richardsons played a role in local politics, whether in office or on the sidelines, but primarily as wheelers and dealers in the political machine. Just keeping their land claim through British, Spanish, and later American rule required political sleights of hand, so the Richardsons were no strangers to backroom deals and crafty if legal maneuvers.

Though Jim worked politics largely through the back door, he began to participate more openly when in the early 1940s he met a young man by the name of James Elisha Folsom. Standing six feet eight inches tall and weighing 275 pounds, James "Big Jim" Folsom towered over everyone around him. Though Jim Folsom would not win his bid for governor in 1942, he and Jim Richardson bonded over his love of Jim's whiskey, but most important his stand on race, which at the time was considered progressive. Folsom and members of his populist movement in Alabama focused more on agriculture and economics than on racial issues. Also, Folsom claimed in stories he told privately along the campaign trail that his family had opposed secession and refused to fight for the preservation of slavery.

These stories were probably not true, but they would have appealed to Jim Richardson, and I am sure Big Jim knew that. Folsom's biographer George Sims notes, "He [Folsom] did not even view segregation as an inherent violation of the Negro's civil rights." Still, as a candidate for governor, Folsom expressed a concern for civil rights, particularly calling for the abolition of the poll tax, which was unfair to blacks and poor whites. Though he never publicly argued that the poll tax was unfair to blacks, this stand would have appealed to Jim Richardson, who had influence among whites and blacks. So, when Folsom ran for governor again in 1946, Jim Richardson worked his network of the powerful men he kept supplied with whiskey to help Folsom carry Washington County.

Jim's involvement with Big Jim Folsom served to protect his local interests rather than reflecting any larger ambitions he may have had. For much of his life, Jim had worked his way through and around the local political landscape. With Folsom, he had begun to think about an interest even closer to home: his children's future. More than ten years before he met Jim Folsom, he had promised Edna on her deathbed that he would take care of their children. Now that they were beginning to grow up and move into the wider world, Jim chose to ally himself with a politician who he thought would make their lives less difficult.

To understand this, I had to go back to November 1929 and the circumstances that followed Edna's death. The Richardson family had always lived their life out loud and in the open. Jim Richardson may have kept the secrets and confidences of other men, but he chose not to live his life wrapped in a secret. Yet, with Edna's death, Jim had to decide how to raise their young children. In the end, I realized that it was Edna's memory that influenced Jim's actions immediately after her death and even up to his death nearly twenty years later.

Life is to be lived, not controlled; and humanity is won by continuing to play in face of certain defeat.

—RALPH ELLISON, *INVISIBLE MAN*

5.

Out from the Shadows

JIM AND EDNA'S WORLD WAS ONE THAT STRIVED TO TRAN-
scend race, but only within proscribed boundaries. They lived
outside society's racial rules as well as within them. But somehow they instinctively understood that identity can be a trap as
much as a source of pride, particularly if that identity requires
concealing elements of one's race in a tightly wrapped secret.
The Richardsons kept no secrets about racial identity, since
masking one's identity could quickly become its own prison,
creating emotions that swell into a realm of the unknown: confusion, self-hatred, fear of being found out. This made all the
difference for the family, particularly as the children grew up.
Their acceptance of themselves led them to feel secure in their
black identity, without rendering them inferior to their father's
identity and status as a white man in the South. It was a critical
choice made for both personal and emotional survival.

Consequently, Jim and Edna lived openly and as normally
as possible, as part of the community's life rather than hidden away at the end of the road. The setting of their house
provided safety rather than isolation. They were regularly

seen together at the train depot in Prestwick, which from the early 1900s up until after World War II stood at the center of the social life in the town. With a general store, the post office, and the regular arrivals of the Southern Railroad, the depot was the place where people congregated, gossiped, and connected with each other. On Saturday evenings and Sundays after church at Mount Moriah, the local AME church, groups would gather at the train depot for what was informally a weekly town social. While chatting and exchanging stories, the group would see who got on and off the train as it passed through Prestwick, providing both fodder for more conversation and, depending who got off the train, additional participants to the gathering.

Some evenings, Jim Richardson would join the gathering, alone on horseback, to talk and socialize. Although he was white and the group was all black, like the town of Prestwick itself, no one thought anything about it. The conversations would keep their usual tone and cadence; the group did not feel the need to show deference to the presence of a white man, as would have been the custom in many black-white social interactions at the time. No one thought of him as someone whose white skin made him superior. He was just part of the community, their community, and that shaped the social dynamic more than race.

Edna Howell Richardson certainly paved the way for Jim to forge these types of relationships in the broader community. In addition to her beauty, she is remembered for her kindness

and generosity, which endeared her to people and in turn soft-
ened Jim's more hard-driving persona, making him more ap-
proachable. "Miss Edna was not a person whose status of being
married to a white man of means made her better than anyone
else," Bernice Bolling Williams told me one cool fall afternoon.
Though she now lives in Mississippi, Williams grew up in Prestwick and remains connected to those who still live there. She
remembered that Jim provided well for his family: he built a
house that at the time was considered palatial for a black person
and made sure that his wife and children did not have less than
their white relatives. Given the family's bounty, compared with
the more meager circumstance of much of Prestwick, Edna did
not see herself and her family as being above or apart from the
rest of the town. "Edna had plenty and she didn't mind giving
to those who didn't."

And those who did not have plenty lived up and down the
winding road that led to the Richardson home, as you can discern from the remnants of the modest dwellings that remain.
Though both are long gone, the memories of Jim and Edna's
generosity have lived on, judging from recollections of those
who knew them. If someone could not afford a burial, Jim and
Edna helped out or paid for it. If someone needed food, they fed
them. There were women in the community who helped Edna
with housework or worked in the kitchen keeping Jim's groups
of loggers fed. They were hired help, but they were treated as
friends and some as family, as in the case of Edna's friend Miss
Callie, who Jim cared for through old age and death.

As people, most of them now near the end of their own
lives, share their memories of Jim and Edna I know it's easy
to romanticize the past and two people who have taken on an
almost mythic status in a fading community. Jim and Edna were
not perfect; they were by nature unpredictable and by most accounts short of temper. Still, in spite of their imperfections and,
by contemporary standards, lack of sophistication, they seemed
to recognize the cultural importance of how race enters a child's
consciousness and shapes personal identity. The stories people
tell me of how they lived and interacted in the communities
make it clear that their sensibilities were instinctual rather than
intellectual, which was key to their survival.

Prestwick fell outside the norm of Alabama and the American South, where there was usually a sharply defined and proscribed concept of race. In an environment like that, it would
not take long for a young person to internalize the prevailing
racial dynamic of white superiority and black inferiority. But
the unique environment created by the Richardsons and the
broader community of Prestwick allowed the Richardson children to escape the full impact of this force. They saw race as a
cultural reality, but not one that defined everything they did.
Judging from my conversations with the surviving members of
the Prestwick community, Jim and Edna worked within their
family and the community to help their children feel secure as
black people while at the same time to not feel torn between
the two worlds their parents represented. Even more important,
they did not want their children to feel inferior to their white

relations. That meant that racial differences were not stressed at home, and the children witnessed their parents treating blacks and whites equally, particularly family members. The only exception to that I have found notable was Jim's mother, Etta.

Although Etta did not approve of Jim's marriage, Edna encouraged her to see her grandchildren, wanting the children to know both sides of the family. Reluctantly, Etta agreed, though on her on terms. "Mama would send us outside to talk to her, but she would never get out of the car. She would just sit in the car, talk to us while we played, but we never had any real physical contact with her," my Aunt Mary remembers. Did Aunt Mary think it was strange? "No, that's just the way it was. I did not question it."

As my aunt's memory makes clear, Jim and Edna did not have a proscribed plan of action for dealing with the racial issues that inevitably arose, even within the family. They took life as it came and dealt with things simply and day by day. In a black community, the influence of Jim Crow laws did not control their day-to-day existence. However, when they left Prestwick to go to nearby Jackson or Mobile, they did encounter the harsh world of segregation. Nonetheless, the family ignored or disregarded Jim Crow laws, and the children were able to do so easily since their outward appearance gave no clues to their being black. For most of the Richardson children, they were well into adulthood until the rules of a segregated society began to have an impact on their lives. Only after marrying my dark-skinned father did my mother, at the age of twenty-three, learn that black people could not try on clothes in department stores.

The real test of the family's life together came about when Edna died in 1937. Her death left Jim with four young children to raise to adulthood as well as some hard decisions to make, from the standpoint of family relationships and the way the family was viewed according to Alabama law.

"I still remember when Aunt Jannie arrived with the empty suitcases to take us away with her, soon after Mama died," my Aunt Mary recalled in a tone so casual that it masked what must have been a traumatic moment. Not long after that sunless November day in 1937, her sister, Jannie Howell Davis, arrived unannounced to take the two younger girls to live with her briefly; now she wanted to make it permanent. Jannie thought it would be best to bring up Mary and Lucille as her own, to give them what she thought was a more stable life than Jim Richardson, with his rough-and-tumble lifestyle of logging and bootlegging, could offer them. In Mobile, the girls would have more opportunities than they would in the woods of Prestwick. She had no interest in the two young boys, who were twelve and fourteen at the time and clearly allied with their father. Jim would certainly get them into his logging business, and they were at an age where she thought she could exert little influence. But the girls were another story. How were two young girls going to survive in this male-dominated household, much less one belonging to a white man?

Somehow Jim had to talk Jannie out of taking the girls

away, and I'm certain it wasn't easy. My mother and her sister don't know exactly what happened, but within a few hours, their aunt took the empty suitcases back to Mobile. Jannie was just as strong-minded as Jim, so there must have been a tense encounter, with legal as well as social issues falling on Jannie's side rather than Jim's. Jannie knew that because the children were of mixed race Jim had no obligation according to the laws of the state of Alabama to provide for the children. According to the law, he could just walk away at any time, without any consequences whatsoever. In fact, the state would have preferred that he abandon his family rather than run the risk of the family possibly becoming white and perhaps lead the way to racial impurity. And, in Mobile, as in New Orleans, there was a sizeable population of "Creoles of color" and light-skinned blacks, a world the girls could easily move into and that would give them what many thought of as a social advantage in what was a tightly conscribed world.

I'm certain that Jim had to convince his sister-in-law that he didn't want to split up his family. Jim knew that once his children were separated from him and outside of the unique world of Prestwick at some point they would have to deny their tie to him, something they would be less likely to do if he kept them together. Plus, he had sworn that he would never abandon his children. He'd faced this choice of giving up his children once before when they were younger, more malleable, and he refused it then. He wasn't going to do it now.

When all of the children were younger, Jim's fishing partner, Dr. James Bedsole, offered to send them away to pass for white. This would have included the entire family. Exactly what the arrangement would have been is unclear, but it did involve Jim losing a direct tie to his extended family in Washington County, who could easily unmask the family's new identity. All the children had birth certificates declaring them as white, they looked white, and given the times, would have had more opportunities had they lived as white people. Jim refused then, because he wanted his children to "face up to who they were," as Dr. Bedsole recounted to my mother before she went away to college at Tuskegee Institute. Dr. Bedsole thought my mother should know this piece of her personal history, particularly now that at Tuskegee she would undoubtedly become deeply rooted in the black community.

After Edna's death, Dr. Bedsole made the offer yet again, but only for the younger children, who were still at a stage in their life where they would be able to keep their background hidden. Given that the children had recently lost their mother, grief would have acted as a shield against anyone who would have questioned their racial identity. Still, Jim's answer was the same as before: he refused.

With Dr. Bedsole's offer, the children would have been raised to hide their identity. Jannie's offer was the same, only different. In order to live in Mobile, the children would have to deny a tie to their white father. Jim knew that time would eventually break down their racial secrets and cause long pent-up feelings and emotions to erupt. Jannie's offer would wrest away his two small daughters, the only ties with Edna that remained. Jim sent Jannie back to Mobile, and life for the family moved on.

ISOLATION CAN SUSTAIN A COMMUNITY AND ITS WAY OF LIFE or wipe it away. As long as Prestwick remained a remote island of racial tolerance in southern Alabama, sustained by timber, a sawmill, and farming, the town and its people flourished. Jim Richardson helped keep the community going because it served both his personal and financial interests. Few others in the community had anything like the resources he did, making it incumbent upon him to be generous so that the people who made the community what it was remained there and provided his family safety and financial security.

Looking at it from the outside, with little knowledge of the place and its history, it's reasonable to conclude that Jim Richardson was a paternalistic white man who kept Prestwick going as along as he could, as if it were his own personal plantation. He owned over 400 acres of land and held considerable political influence, unlike anything a black man could have had in this part of Alabama in the first half of the twentieth century. People looked to him for leadership and to help sustain the meager infrastructure of the town. But that infrastructure—the store, school, and church—were controlled by black people. To his credit, Jim did not seek to control those institutions as a plantation owner would have.

According to the people who remain in Prestwick and remember a vital, strong community, Jim Richardson did not function as the master of Prestwick and its people. He was a part of the town, and he was one of them, although there was no doubt that his influence stretched beyond its tiny borders. "I met Jim Richardson when I was seventeen years old and my brother worked logging with him," Shed Gregg recalled. He was ninety-five when I talked with him, with memories of old Prestwick undimmed by his age. "When I met him I approached him the way my people taught me to act around white people," meaning that he was both cautious and deferential. "But when I got to know him, I stopped thinking of him as a white man. He was just a man and I trusted him. I respected him because he shared what he had with folks who did not have as much as he did and he provided for his family. And he remained my friend and returned that friendship." Shed's relationship with Jim grew over the years and became stronger toward the end of his life. One incident just five years before Jim died sealed their bond and let Shed know that Jim would do anything for his family. It also gave Jim a taste of what his children's life was like on the other side of the color line.

In December of 1951, Jim's son Robert, known as "Smokey," was caught by an FBI agent in a raid on his liquor still. At the time bootleggers produced more liquor than all the legal U.S. distilleries combined, making it a large underground, untaxed economy.

The agent, Grover C. Trumbo, was part of an FBI effort to

crack down on the tax-evading bootleggers, but he was acting alone when he arrested Smokey and an associate. Newspaper accounts say Smokey asked Agent Trumbo to stop at his house so that he could tell his wife and children that he was going to jail and "Trumbo allowed the colored man to go into the house alone," the *Mobile Register* reported. Smokey emerged from the house brandishing a shotgun, fired at, and wounded Trumbo. He then shot out the tires of Trumbo's truck and fled into the woods.

Jim caught word of the incident when the entire local police force and the highway patrol converged on Prestwick. While Smokey hid in woods that he knew better than any of his pursuers, Jim began to negotiate with the police, who by Shed's and Smokey's recollection of the event, wanted to kill Smokey, just as they would with any black man who shot a white man. By this time, Jim was not in the best of health, but he put all of his energies and resources into saving Smokey's life. "I won't turn him in unless you promise me you won't hurt my boy," Shed remembered Jim saying to the authorities, even though his boy was a man of thirty at the time. It took three months to negotiate turning Smokey over to the police and FBI and for him to be tried, convicted, and sent to federal prison for three years. Both Shed and Smokey agree that had Jim not been a white man who had influence in Washington County, with a network of lawyers and influential politicians to draw on for assistance, Smokey would have been hunted down and killed.

With a blurry microfilm copy of the newspaper article in

hand, I asked my Uncle Smokey if he was sure the police were going to kill him. Given my uncle's ability to tell a good story, I wanted to do my best to separate fact from the folklore that has built up around this incident. "After all," I reminded him, "you did shoot an FBI agent." Looking at the article, he points out that the paper makes it clear that he is "colored." "Colored men didn't shoot white men back then without white folks coming to shoot you down."

My uncle's statement made me see the story in a new light. What he did was against the law, but in the Jim Crow South, laws often had nothing to do with how a criminal was treated. A black man committing the same crime as a white man in the South could expect to be treated differently, since blacks were rarely treated equally with whites under existing laws. And as I explored the reports surrounding my uncle's arrest, I found hints of some of the discrimination that led my grandfather to worry that my uncle would have been harmed had he immediately turned himself in.

While reading the documents around my uncle's arrest, one of the first things that caught my attention was this description of him: "Richardson is the son of Jim Richardson of Washington County Alabama. Richardson's father is a white man, and his mother is a negress." The use of such a pejorative term to refer to my grandmother, one that denotes that she was mere property rather than Jim's wife, shocked me like a slap in the face. Right away, I saw that a racial line had been set so that those pursuing my uncle would know that despite his outward appearance, they were after a black man. For his race, police and FBI documents identify him as "Negro (half white)."

In interviews with local authorities, the FBI agent noted that the deputy sheriff is "well acquainted" with my grandfather and knows that he has "a firm respect for the law and has in the past on a number of occasions produced his sons voluntarily to the sheriff when they were in trouble." I knew this was true, since I learned that my grandfather even kept an attorney on retainer for my uncles' run-ins with the law, whether it involved bootlegging or gun play. But this time, Jim Richardson was not willing to turn in his son. In this case he did not trust the authorities and seemed to be purposely vague in his communication. "Smokey could be in France, Mexico, or even up North. I have no idea of where he is," was his response to the investigators, though I'm certain Jim knew that Smokey was in and around the swamps surrounding Bolen's Lake, right near his house. In the middle of the lake was a small island and a fishing shack, and since my uncle knew the swamps and woods around those parts better than those pursuing him, he was able to keep himself hidden. In March 1952, almost four months after the shooting, my grandfather got my uncle to surrender to the deputy sheriff. Somehow, Jim got every assurance from the deputy sheriff that my uncle would not be harmed, but as always, he left nothing to chance. To make sure that nothing changed from the deal he had brokered, Jim accompanied my uncle to Mobile, where he paid a $5,000 bond and then drove my uncle home himself.

According to my uncle, this whole incident made it painfully clear to Jim that his children's circumstances as black
people did not afford them the same treatment as white people.
He knew it before, but this experience drove the issue home.
Soon after Smokey's run in with the FBI, a weakening heart
forced Jim to slow down his life. He realized that the way of life
his family had had in Prestwick would probably go to the grave
with him. With his usual shrewdness and cunning he started
to plan for life after he was gone. And he knew that although
his family and friends included black people, there were white
members of his family who did not believe that the assets he
built over his lifetime should fall into black hands. "I know that
when I close my eyes, my family will try to take everything
from them," Shed's wife Vivian recalled Jim saying when he
knew he did not have long to live. Jim started to deed all the
property he had to his children, since according to Alabama
law, the children from his marriage to Edna were entitled to
nothing since there could be no legitimate children from an interracial union. Even if he spelled out in his will what they were
to have, legally his family could contest it and take everything
from his children. The law would have been on their side.

Race is one of the most powerful dividing forces between
people, even people within the same family. In the South, for
interracial families like the Richardsons, family was defined by
race rather than blood. Two families could have the same last
name, fathers, uncles, and cousins in common and the taint of
black blood would irrevocably separate them, making them dis-

tinct entities rather than part of the same family. Consequently, Jim and his immediate family were largely invisible to their relatives who lived in neighboring Chatom.

In life and in death, race divided the Richardson family. Yet while Jim lived and after his death, there were members of his white family who maintained family ties with him and his family. As a boy, I remember a steady stream of fair, blue-eyed people coming to see my mother when we would visit Prestwick. Because they looked like her, and my mother was black, I thought they were black like her. They were not, but they were family nonetheless and conducted themselves in an affable manner that made all of the children see them as family regardless. Now I know that there are those who did not want to sustain those family ties and even worked to bring about disunion between those Richardsons who identified as black and those who were white, even though if you stood them all together they all had the same coloring and features. Those were not the people I encountered in Prestwick, because they had not succumbed to the thinking of broader white Southern society where separation of the races was insisted upon as the norm.

When Jim died in December 1956, the struggle between the black and white families began to play out on a bigger stage. After Jim's death, his sisters Claudia and Sadie asked to have his body buried in the Richardson family cemetery. Claudia had maintained a relationship with Jim over the years, as had Sadie, but the relationships were different. Claudia had remained in nearby Leroy, near where the siblings had grown up, and she always included Jim's children as part of the family. Sadie had moved to Mobile and become a part of its society, embracing the norms of the segregated South. She wanted nothing to do with Jim's children and would leave the room without speaking when they came to see him in the hospital near the end of his life. "We had him all of those years, so we decided that if they wanted him now, they could have him back," my mother always told me. The decision to bury Jim in all-white Pine Grove was left to her, since as the youngest child she had the closest relationship with her father. She begrudgingly agreed, and a graveside service with the entire family, both black and white, was held at Pine Grove.

When I visit Prestwick, I often stop at Pine Grove to pay my respects to Jim. Each time I walk through the cemetery gates and stop at the grave, I pause in rapt wonder and think, sometimes overcome with puzzlement, "who were these people who wanted him back?" After piecing together Jim and Edna's life in what remains of Prestwick, I decided that the next thing I needed to do was to find out more about the people who wanted nothing or little to do with him in life, but wanted to take him back in death. I had to know more about these people because, for better or for worse, they are my family.

REACHING ACROSS THE CHASM

Draw a dead line between the races. Tell the Negro, when he crosses it the penalty is death. Tell the white man, when he crosses it, the penitentiary is there.

—GEORGIA JUDGE THOMAS M. NORWOOD, "ADDRESS ON THE NEGRO," 1907

6.

A Beautiful, Needful Thing

ON BOTH SIDES OF MY FAMILY, I HAVE A GRANDPARENT with blond hair and blue eyes. If I had photographs of the two of them placed side by side, you could not tell which one was white and which one was black. In their unique ways, both of them blurred the lines of identity dictated by the state of Alabama by virtue of their outward appearance and their personal sensibilities. In the case of my mother's father, Jim Richardson, some in his family and the community at large believed he betrayed his race and background by marrying a black woman. Somehow he crossed over to being black by association. Nonetheless, he moved fluidly between the black and white worlds with little regard to the boundaries established by either group.

My paternal grandmother, Sally Young Eubanks, was confident in her black racial identity and took pride in it, despite an outward appearance that would have allowed her to move into the white world alongside Jim Richardson. Unlike my grandfather, an element of Sally Eubanks's background froze her identity firmly in place: she was the daughter of a woman who had been born a slave. Her outward appearance came from the secret history of interracial liaisons in the South, with her background being something that was whispered about and purposely forgotten. People in my father's family sometimes spoke softly about a link to a prominent Alabama political family, but I've never known whether this is true nor have I found any means to confirm it. There is no traceable family tie, and that's exactly the way it was intended. Histories of liaisons such as those that produced my paternal grandmother were kept in dark corners where no one could discover them and discuss them openly.

In the South, the era of slavery established both the pattern for interracial relationships and led to laws prohibiting those relationships. Tangled up in laws governing slavery were guidelines that tightly demarcated racial identity, defining as black anyone who had a blood tie to a slave, despite any outward appearance to the contrary, to keep all black people enslaved. When slavery was abolished, race and color replaced slavery as part of a classification system for oppression. Some system had to replace the social order destroyed by the abolition of slavery.

The state of Alabama saw its primary responsibility to protect white families, and interracial families were viewed as a direct threat to white families, as well as white supremacy. By the time of Alabama's constitutional convention in 1901, white supremacy became not just an idea, but a political tool. In his opening remarks to the convention, the convention's president

John Knox noted, "If we would have white supremacy, we must establish it by law—not by force or fraud." In Alabama's new racial hierarchy, illicit interracial relationships were acceptable. Interracial marriage was not.

As long as interracial relationships were kept secret, and no one had to acknowledge them, they were accepted with a wink and a nod. But when they were open and might be recognized as legitimate marriages, as was the case with Jim and Edna Richardson, the rules of the game changed. It was the fear of legitimizing interracial marriages, and the children resulting from those marriages who might move into the white world, that led to strict laws such as those in Alabama that prohibited marriage across racial lines and established criteria for who could legally say they were white and who was legally black.

THE DESIRE FOR FREEDOM IS CENTRAL TO HUMAN NATURE, alongside the desire to love and be loved. Love and freedom help us feel and embrace the beauty that surrounds us, whether in people or where we live. Our very humanity depends on these two beautiful, needful things. But when freedom is swept away by laws or ignorance, humanity and love are cast aside in the process.

It's painful to read about love and the freedom to love being punishable by death or imprisonment, reducing something beautiful to something ugly. Though it's hard to imagine today, the risk of death and imprisonment stood as a constant backdrop to my grandparents' life as a married couple. In Alabama and throughout the South, prosecution for interracial marriage served as a means of defining race and racial identity as well as establishing white supremacy. Up until the Civil War, slavery divided the races and established who was subordinate or inferior. With slavery abolished, something had to replace that racial dividing line and dictate strict societal roles: who could marry, who could not; who was black and who was white. As a result, everything in my grandparents' lives, from legal arrangements for the ownership of land to the place they lived, was constructed to protect them from forces that dictated that their marriage and family existed in violation of the law.

But somehow, Jim and Edna Richardson weaved their way around the law. Intermarriage, in the eyes of most whites, meant a black man marrying a white woman, yet the law was clear: all intermarriage was illegal and ran the risk of prosecution since it violated prevailing social norms. How did Jim and Edna avoid getting the law entangled in their family life? First, Prestwick was remote and functioned as one of Alabama's independent "island communities," which was probably part of their first shield of defense. I'm told that Jim and Edna took great care in how they publicly characterized their relationship, since anything they said could have been used as a confession in a miscegenation case. Also, Jim Richardson's well-known temper and facility with a gun scared any would-be prosecutors out of the way. Add to that Jim's sideline as a bootlegger, and a more pointed question arises: why wasn't he snared for illegal liquor sales as a means of punishing him for crossing the color line with his marriage? Jim sold whiskey to the powerful and prominent men of the county, some who could have prosecuted him, but may not have for fear of what he knew they had concealed in the shadows of their own lives. Whatever it was that kept my grandparents safe from the ugliness of Alabama's miscegenation laws, I decided to dive headfirst into any archival sources I could find about intermarriage in Alabama to figure it out. I began to explore legal cases where couples were prosecuted for intermarriage. Within these cases, I looked for circumstances that matched those of my grandparents.

From the late nineteenth century up through the 1930s, Alabama enacted a confusing swirl of laws prohibiting interracial marriage and constructed arcane legal definitions of racial identity. On top of that, Alabama courts issued more court decisions on intermarriage than any other state: thirty-eight opinions between the end of the Civil War and the invalidation of laws against intermarriage in 1967 by the U. S. Supreme Court. Although this is not a large number of cases, it is significant because the number is larger than those in other states, a sign of how vigorously the state viewed the regulation of interracial marriages. Not only did Alabama's courts defend laws against interracial marriage on the basis of how interracial relationships affected the fabric of the state, but also on how they affected the fabric of families, which is viewed as a basic unit of the state. So, one of the purposes of regulating interracial marriage, as outlined in one case, was to prevent "the evil of introducing into their most intimate relations, elements so heterogeneous that they must naturally cause discord, shame, disruption of family circles, and estrangement." In the eyes of the law, an interracial family was not only abnormal and illegal, but also evil.

The work of Julie Novkov, a legal scholar specializing in the history of miscegenation laws, served as my guide through this confusing net of cases and laws. Her detailed analysis of the body of Alabama's miscegenation cases pointed me to stories that aligned with the life of my grandparents, whether they happened near Prestwick or in similarly isolated areas, or even if one party to the case was a bootlegger much like Jim. There was a bootlegging case, but the details of the case revealed that this bootlegger lacked Jim's cunning. There were cases that involved white men and black women, which defied my thesis that my grandparents were not prosecuted because the state of Alabama only viewed relationships between black men and white women as socially disruptive and violations of the principles of white supremacy. Still, many of the cases involved black men and white women, combined with racial violence and intimidation. Few of the cases were in south Alabama, leading me to wonder if miscegenation cases were more likely to happen in other regions that did not have large pockets of people of mixed race, like Washington County. Soon, one case in particular whose time and place paralleled my grandparents' lifetime drew my attention.

In 1915, about the same time as Jim Richardson and Edna Howell married, another couple in Washington County, Percy

Reed and Helen Corkins, also began living together as husband and wife. After they had been married for about five years, they were charged in Washington County with the crime of intermarrying illegally. At their trial, the focus was not on Percy Reed's race, but on the race of three generations of his ancestors. This approach was taken because, by 1920 in Alabama, race was not legally defined by outward appearance but by blood. To prove the crime of illegal intermarriage, the testimony of witnesses to the jury had to show beyond a reasonable doubt that someone, anyone, in Percy Reed's lineage was black.

Bear in mind that well into the twentieth century, race was a scientific term. Medical doctors, anthropologists, and physiologists together developed the scientific idea of race, one based on blood rather than earlier more amorphous notions that there were physical and moral differences between races. The end result was the same: whites were superior and blacks were inferior beings. At the time of Percy Reed's trial, eugenics had also come to play a role in how race was legally defined in Alabama and other Southern states. According to the social philosophy of eugenics, racial mixing resulted in inferior human beings, since European whites were biologically and culturally superior to blacks or other racial and ethnic groups. This meant that black blood was dangerous: the inferior characteristics of blackness could be passed on from one generation to another, thus corrupting the intelligence and refinement believed to be central to whiteness. Heredity was destiny. The only way to stop the amalgamation of the races, which the proponents of eugenics believed would contribute to the decline of American civilization, was to have strict prohibitions protecting the integrity of white blood. So, in the interest of keeping Alabama society as racially pure as possible, Percy Reed was brought to trial. Reed would be in violation of laws against interracial marriage if any of his ancestors were found to be black. Consequently, the trial did not focus on his race but on the race of his great-grandmother and her daughter, Reed's grandmother. In framing the case for the jury, the judge asked the members of the jury to consider whether anyone in Reed's lineage was a "full-blooded Negro." If he had a full-blooded black relative somewhere in his background, Percy Reed would face up to seven years in the state penitentiary.

The case drew me in because of the familiarity of the name, Reed. Then I remembered the story of Daniel and Rose Reed. Daniel Reed, a freed slave from Santo Domingo, had purchased a light-skinned woman named Rose from her owner, Young Gaines. The bill of sale for Rose Reed was among one of the first documents I had found in the Washington County Courthouse as I explored the interracial origins of the place my grandparents chose to call home. Interestingly enough, that bill of sale never turned up in this trial. To prove that Percy Reed was black, the state of Alabama sought out sworn testimony from people who knew Percy Reed and whether he had any black relatives.

At the trial, two men testified on behalf of the state of Alabama against Percy Reed. In a sworn affidavit, one man, attesting to the fact that Reed's great-grandmother was black, noted that she was a "ginger cake color," probably a mulatto, with her daughter, Reed's grandmother, being "brighter" in skin color. The other man testified that he knew Percy Reed was a mulatto because he was "moxed [sic] with Negro—have been told that, and I also judge from his looks."

The first man who testified at the trial was John Richardson and the second was A. G. Richardson. These two men were close family members of my grandfather's: Jim Richardson's uncles, his father's brothers. This discovery only added to my puzzlement, since the same Alabama law they sought to provide testimony in support of was being ignored by their own nephew. Were their actions in this trial to make up for their family's interracial history? Based on what I had uncovered through oral history, I knew that the whispered gossip and stories about Jim Richardson had diminished the social standing of the family in the eyes of some in Washington County. This action by Jim's uncles bears the marks of a hasty grasp at social redemption.

Then, another thought occurred to me: John and A. G. Richardson also had to prove to the broad society of Washington County that they and their families were racially pure. To do this, they had to demonstrate that they were upholding one of the cornerstones of the time and society in which they lived: white racial purity. Based on prevailing cultural notions at the time, these two men were compelled to testify in this case to prove to others in the community that they did not condone the behavior of their nephew, who, by virtue of his marriage and living openly with a black woman, had betrayed his race and sullied his identity and standing as a white man. Jim Richardson, as far as some members of his family were concerned, had crossed from one racial identity to another and sometimes back again. This meant that his children could follow his lead and cross into whiteness just as he had crossed into blackness. John and A. G. Richardson's involvement in the Reed trial sent the message that amorphous or ambiguous racial identities would not be tolerated within the Richardson family.

At the time, around 1920, most Southern whites believed that impurity of the races violated the laws of nature. In the view of some whites, interracial relationships and the children that came from those relationships were akin to incest. If John and A. G. Richardson had condoned interracial relationships, whether in their family or in Percy Reed's, they would have diminished their social and political standing by ostensibly accepting a practice that was thought to be unnatural. Plus, they would have called into question the prevailing standards establishing racial identity, leaving little or no dividing line between black and white.

All the Richardsons in this large extended family were white, white meaning that they had no known black ancestors. Alabama legally defined race based on heredity or a "one drop" rule: if you had any degree of black ancestry, you were black. By testifying in this trial, John and A. G. Richardson were attesting to the fact that their family was pure, untainted by a single drop of black blood.

Today, the one-drop rule still influences American culture's view of race and racial identity, even in some segments of black America. In the American South, including my home state of Mississippi, a racial designation is still required in many states to obtain a driver's license. Consequently, at a certain age, a multiracial child is forced to choose a racial identity, a decision often steered by the one-drop rule rather than personal choice or self-identification.

The Richardsons of Prestwick unwittingly adhered to the one-drop rule. By social convention, both by virtue of living in a black community and the standards of the dominant white culture, they were marked as being tainted. In the eyes of the white extended family, they were not viewed as part of the same family legally or culturally.

The one-drop rule divided the Richardson family and continues to keep many of those divisions intact. Both families have the same surname and relatives in common. Yet these two sides of the family largely live separate lives and do not think of themselves as one family. They are separate families, with the descriptor of "black" or "white" prefacing the surname so that you know exactly which Richardsons are being spoken of in a conversation.

The case of the state of Alabama against Percy Reed ended up being damaged by testimony that many members of Reed's family had intermarried with local white families. One witness identified four white families with whom the Reeds had inter-married, and another claimed that they had established blood ties with "two thirds of the people in that part of the county." This meant that there might even be members of the jury who would have a blood tie to the Reeds. Still, Reed was sentenced to three to four years in prison, only later to have his conviction overturned.

I am one of those who John and A. G. Richardson would have thought of as tainted, in spite of our direct familial link. Undeniably, there are physical similarities that would link us as family, which I can see when I glance at pictures of these two men, my grandfather, and my mother and her siblings. But the markers for familial and racial authenticity have kept most members of these two sides of the family separate for two generations.

The actions of John and A. G. Richardson linked their family legacy to a system that constructed racial identity purely for the purpose of oppression. Although it has been nearly a century since the trial of Percy Reed, I still wonder how much of the attitudes of these two Richardson men live on in their descendants. Do their grandchildren and great-grandchildren know of this legacy? Does it still play a role in their lives? What have they done to reconcile themselves with their ancestors' mistakes?

Did my grandfather's uncles have the same mindset as the members of my family who wanted my grandfather back when he died, to make sure he was buried in a white cemetery and would not be labeled black by virtue of where he was buried? Two actions, though entirely separate, have become linked, leading me to wonder how much these two events from the past reverberate in the present. Have the Richardsons and their descendants moved beyond this tangled world of racial division or have they internalized parts of it? I began to seek out these Richardsons who have remained separate from those I grew up with, as well as the descendants of my grandfather's siblings, to see if they understand the same connections to the past that I have made. Will they think of me as family? That is how I have come to think of them, in spite of past actions. Or will they reject me or deny that the actions of the past have any meaning in the present? Most important of all, I wonder if they can also comprehend the beautiful, needful thing that Jim and Edna Richardson were trying to embrace through the way they lived their lives.

Ambiguous of race they stand,
By one disowned, scorned of another,
Not knowing where to stretch a hand,
And cry, "My sister" or "My brother."

—COUNTEE CULLEN, "NEAR WHITE"

7.

Parallel Lives, Separate Legacies

BECAUSE OF THEIR RACIAL DIFFERENCE, THE LAW DICTATED that the black Richardsons and white Richardsons were separate and distinct family entities, not part of the same family. Again, bear in mind that by law, the children of Jim and Edna Richardson were illegitimate, making them people who did not have to be recognized as family. By social custom, these black Richardsons were inferior, and that inferiority divided the two branches of the family. Now, with those laws and social structures cast aside during the life of several generations of descendents, it is time to look at the family through what unifies these two branches rather than what at one time divided it. And what unifies it is a common ancestry and history rooted in the origins of the Alabama frontier. Interracial relationships were a part of that frontier tradition, making the Richardsons no different from other families in the region. A shared history of straddling the racial divide might be used as a means of unifying the family rather than dividing it.

There is much that can be interpreted from the life of Jim and Edna Richardson, but there is also much that can never be known. Lingering in the few scraps of writings and legal documents they left behind are clues and traces of who they were and what their life together was. Like all marriages, theirs was not perfect. It was filled with compromises, the biggest being the house Jim built for Edna to draw her back home after she left. But wrapped in all they did for each other, both big and small, were acts of love made unwaveringly in a world that did all it could to make life difficult. In spite of the laws that declared the family illegal and sought to fracture their existence, they became a family.

Jim and Edna's burial places are demarcated by race, the very boundary they broke with their lives. Yet in death they were not divided, since their unified sense of courage lives on in the memories of their children and those who remain along the road leading to their house. By examining their demonstrated courage and headstrong determination to defy the forces of white supremacy I have been able to interpret pieces of their lives and come to know them as best one can from someone who cannot speak and tell their story.

Combined with the stories that remain in the memories of those who knew them in life, these tiny bits of information disclose components of their personality and spirit yet leave other aspects of their inner life concealed. Perhaps that is what they wanted. What matters most is that they lived life in good faith, with little regard for what their legacy would be or how others would think of them after they departed from the sandy soil of

Prestwick. The detritus of their life that survives only allows us to know them imperfectly. Yet it is within those imperfections that their humanity shines through.

Within those imperfections stands a sense of family rooted in desire and need. It's much more chaotic and discordant than my own sense of family; the times they lived in, combined with legal dodges and calls to shift their identity, simply made their family that way. Yet it is a family that resounds in life and memory in spite of forces that would have wrenched it apart. Some of those divisive forces came from society; some existed within the extended Richardson family.

Much has been written in literature, law, and science about who is black and who is white, and custom combined with law defined and complicated racial divisions and divided families by race. Mathematical formulas were even constructed to determine how much of a person's ancestry was black or white. A mulatto has one black and one white parent; a quadroon has one white and one mulatto parent; and an octoroon has one white and one quadroon parent, and so on. In speaking of my Richardson relatives, I often refer to them ambiguously as people of "mixed race," feeling compelled to explain why they are fair of skin tone even though they would only speak of themselves as being black. But black or African American is what and who they feel they are inside, and it is the identity that has proudly been passed down on my side of the family.

Identity does not come about through formulas and charts,

but it develops within a family and community structure. And sometimes a community forces one identity or another. Among the Richardsons, there are several different social and cultural identities, many of them with their origins in race and some deriving in local customs or traditions that flowed out of the community. Although he would not deny being black, my Uncle Edward held two driver's licenses, each with a different racial designation. When asked which one he used, his reply was "whichever one is convenient under the circumstances."

As I began to look beyond Prestwick and the branch of the Richardson family I grew up knowing, I discovered some parallels among the various branches of the family and their identities. At the same time, I learned of the fissures that divided them, both then and now. There were divisions that I did not expect to find, as well as some that were expected. What I found was the two sides of the family had more two-way traffic between them than I thought. Often, divisions were dictated by the broader culture rather than by the individual family members themselves.

DEGREES OF SEPARATION VARY WITHIN INTERRACIAL FAMIlies, and the Richardsons are no different in that regard. To make connections across the family's racial divide, I began with those who were accepting, even in the most minimal of ways. I sought out family members who, without blinking or casting down their eyes, would acknowledge a familial link. My black cousins told me stories of relatives who would duck or move to the other side of the street if they saw them in public. For these relatives, the door is probably shut tightly from accepting a family member who is not white. To get the door open even a crack with this side of the family, I did not begin by knocking on their doors. There would have to be some gentle means of introduction for a connection to be made. Then I remembered that there were white relatives who would come to see my mother whenever we visited Prestwick. I did not realize they were white at the time. For those who grew up in close proximity to black family members, there must have been acceptance of the difference as well as a real family bond. Just as these members of the family served as my bridge from my grandfather to his less accepting white relatives, I thought they would also reach across the family divide for me. And these family members lived just over in the next community, Leroy.

As highways overtook railroads as the primary means of transportation, Leroy subsumed Prestwick as a town. When my mother was growing up, Prestwick, by virtue of its saw-mill and connection to the Southern Railroad, had more local importance than the little town of Leroy. Prestwick had the train station closest to Leroy, which served to tie these two communities, one black, one mostly white, together. All of that has changed now that the four lanes of U.S. Highway 43 serve as the main route to Mobile rather than the railroad. Leroy is now more central to the life of this part of Washington County, even though it's is not much larger than Prestwick was in its glory days.

So, one afternoon I drove down the winding road from Prestwick to Leroy to make my first family connection with my white relatives. Two names were provided by my Uncle Smokey: his first cousin Woody Richardson and Pat Foster, the granddaughter of Jim's sister Claudia. Woody lived in Leroy, and Pat he was not so sure about, but he thought Woody would know where I could find her. Smokey had always had a good relationship with Woody, so he thought he would talk with me and help make some connections. It seemed like a good place to start.

Woody's business, Leroy Motors, sits on a lonely gravel plot just off the highway, a used car lot dotted with a few aging Fords and Chevys and some cars out for repair in the back. There doesn't seem to be much activity here, in spite of all the traffic that whirs by heading north toward the Tombigbee River, to nearby Clarke County or to Mobile in the south. As I open the front door, I see a man sitting inside in an easy chair. We look at each other, both of us sensing some vague connection. Quickly, I introduce myself and tell him who I am. "You're Lucille's boy, aren't you?" he says jubilantly, which put me at ease, since I feel like a nosy intruder. Soon I discover that he and my mother, being the same age, were once playmates. Jim often took Woody around with him after his father was killed in a car accident. The bond between my grandfather and Woody was strong, with Jim serving as the father Woody never really got to know. It wasn't long before our conversation speeds forward and we begin to discuss the family's racial divide. Without hesitation, Woody tells me where the line is drawn between the two sides of the family.

"People in the family who moved into town, into white social circles, wanted to deny any blood tie to Jim Richardson and his family. If you got into white society, that's just what you did." In white society, outside of Leroy, Prestwick, or Carson, black people had their place and that place never included the family table. Even today, Southern white society is very class conscious, with family pedigrees checked carefully before a newcomer is accepted. But in my grandparents' days, if you chose to associate with blacks and, even worse, view them as family, social doors were closed. In order to open them, you had to deny any close relationships with black people, other than someone subserviently in your employ. You could certainly not acknowledge that you were related to them. And that's how many Richardson relatives behaved, including Jim's uncles John and A. G. Richardson. The social denial dictated by tradition is the real divide in the family. It's not simply a black-white divide; it's a divide between those who lived in the remote area south of Bassett's Creek, where there was a more open, accepting culture with few of the traditional rules of white Southern society. Then, there were those who lived north of Bassett's Creek, in small towns such

as Chatom or communities with few black residents and with
strong social and racial divisions that were the norm in the
South.

Woody told me that there were white Richardsons who
felt pressure from their social network to distance themselves
from their black relatives. One cousin was reminded by several
suitors that they would not marry her because "you're kin to
some niggers." The problem was not just one of having direct
kinship with blacks; the issue was having that kinship and acknowledging it openly. Consequently, this cousin left Leroy, but
others stayed. Jim had one sister, Sadie, who left Washington
County for Mobile so that she would no longer have a link to
her black relatives. But another sister, Claudia, chose to stay,
and she maintained a tie to Jim and his family.

Of course, Claudia died more than twenty years ago, making me remark to Woody that I wish I had started making these
connections well before today. "But if you started then, no one
would have talked to you," he quickly noted. He was right. It
would have been too close to the changes in the social dynamic
wrought by integration for people to speak openly twenty years
ago. Back then, there was anger and frustration about how the
push for equality that came with the civil rights movement had
begun to transform the racial dynamic in the South. There was
also denial that things would continue to change; silence was
how that denial projected itself. A generation later, the old way
of life has been uprooted and the changes of the civil rights

movement are now tenuously ingrained in the social fabric, making it easier for people to talk about the way things once were. Now, I just had to find someone who knew my mother's Aunt Claudia and could tell me more about why she kept a link to Jim and his family.

I kept plying Woody with questions, but he said that he had told me all that he knew. As he stood up from his chair, I noticed a visible tremble in his hand and an overall unsteadiness. "Parkinson's is starting to take hold of me, so I'm not as spry as I used to be," he explained. I mentioned that Smokey thought he would know how to contact Pat Foster, Claudia's granddaughter. Woody gave me her address in Florida. She had left Leroy as a teenager, but returned to visit throughout her life. He was certain that she would talk with me. And Woody's manner and smile revealed a sincere desire to help, just as his physical condition let me know that he had indeed done all he possibly could.

Making contact with Pat Foster was a joyful occasion of connecting with a long lost relative. Because she left Washington County when she was fifteen, she had limited connections with her extended family, but she remembered my grandfather fondly and wanted to fill in pieces of the story I was curious about. Leaving at a young age also gave her some perspective, allowing her to be open in a way that someone who still lived in a small community could not be. "It was always a happy time when Uncle Jim came to visit," she remembered one day in a phone conversation. "My grandmother was a workaholic, but she would stop whatever she was doing and sit with Uncle Jim. He always brought her something to drink, a little toddy made of his homemade liquor."

Perhaps because she was young, she had never heard the issue of race come up about her other relatives, even though she visited with them all on trips back after she moved away. Two of my uncles farmed her grandparents' land and worked side by side with them, yet the issue of race never came up in their conversations. "During that time, there were things people did not talk about in front of children, and race was one of them," she remembered. But she knew that her other relatives were different as evidenced by the way in which the two sides of the family interacted. "I do remember this: I remember thinking that when I was growing up, why did Uncle Smokey come to the back door and call my grandmother 'Miss Claudia' rather than 'Aunt Claudia'? It seemed strange then and seems even stranger to me all these years later," she remarked.

I agreed. Behaving in such a deferential way today, particularly to a relative, would be unheard of. It would have been hard enough to have inferior status dictated by society and even harder to have it dictated and reinforced by a relative. Yet, I imagine that Claudia did not think this nod to Southern racial custom unacceptable. Custom was the cloak that kept racism hidden in the South, in the minds of both whites and blacks.

Until the reforms of the civil rights movement took hold, racial slights such as going to the back door of a white home remained a part of everyday life. A habit, albeit a psychically destructive one.

Yet, within their fragile, socially dictated means of interacting with each other lay an undeniable emotional bond formed in a way that today is difficult to comprehend. Behavior that seems unjustifiable today was part of the complicated social dance whites and blacks had to follow. The genuine familial bond Smokey and Claudia shared had to be masked by racial custom. After speaking with my uncle Smokey, it was clear that neither he nor his aunt liked this means of interaction, but the rest of the world insisted upon it and they dutifully followed the unspoken rules of society. In later years, when Claudia became ill, my uncles Smokey and Edward cared for her until her death. It was then that they closed the distance they once had to put between them.

After a few phone conversations, Pat and I met early one morning near her home in Florida. I had no idea what she looked like, but she immediately picked me out of the crowd of people in the Jacksonville airport. Surprised, I asked her how she found me. "You hold your head just like Uncle Jim, that's how I knew who you were." Then, the image of the portrait of my grandfather came into my mind's eye, and I realized she was right. Family traits are hard to hide, even in a crowded airport.

After she gave me a list of names and phone numbers of other relatives who she thought would help, we talked a bit about my grandfather's other sister, Sadie. While Jim's sister Claudia allowed my aunts and uncles to visit, Sadie would never have acknowledged them or allowed them to even set foot on her back doorstep. Pat did not have any ties to that part of the family, since they distanced themselves to the point of almost complete separation. This fit the pattern Woody had described to me. Pat knew only the names of Sadie's children and where they once lived, since any real family bond had long since disintegrated and her own attempts to find this part of the family had failed. Then Pat told me a bit about her family and the need she feels to have family ties, now that her closest relatives are gone. She also told me about her grandson, whom she is now raising. She pulls out her wallet to show me a photograph of him.

The photograph is of a brown-skinned boy in a baseball uniform, standing proud and strong. His name is Evan, her daughter's son from an interracial marriage. Her daughter and son-in-law have since separated, and now Pat has the responsibility of bringing him up. "Sometimes I'm not sure that as a white woman, I'm the right person to raise him. I want Evan to have ties to my family, but I also want him to be proud of that part of him that's black." I knew what she meant, since the puzzle of race and personal identity will undoubtedly be a part of Evan's life.

At that moment a family bond was formed, free of any

social dictates and rules. What once separated our branches of the family had brought them together again. Before I left to catch my flight, I told Pat reassuringly, "Be sure to tell Evan that lots of members of his family look just like him. Lots of them."

The truth about us, though it must lie all around us every day, is mostly hidden from us, like birds' nests in the woods.

—WENDELL BERRY

8.

A World Lost

WHEN A FAMILY IS DIVIDED, MUCH IS LOST. OLD WOUNDS fester, new ones open, even in spite of mutual silence between the parties. More important, the truth about what led to those divisions becomes more elusive with the passage of time. Throw in the issue of race along with money, property, and personality squabbles, and the combination holds the potential to rise to an unexpected level of volatility. Conversely, the changes to the racial landscape during the past forty years could help transcend differences and transform the way the two sides of the family look at each other, leading to redemption and reconciliation.

All these factors rattled around in my head as I contemplated reaching across the divide in the Richardson family. No matter what might lie in store, the time simply seemed right. Nearly a century had passed since my grandparents' marriage and much had changed, including the removal of the ban on interracial marriage from the state constitution of Alabama. Yet there were signs that not everyone was pleased with this change. Stripping the ban from the constitution was only put to a statewide vote in November 2000, more than thirty years after the U.S. Supreme Court legalized interracial marriage. And though the amendment passed, it was not by as wide a margin as would be expected. Even with 66 percent voter turnout, more than 40 percent of voters rejected it. All in all, this meant that half a million people voted in favor of keeping the ban on interracial marriage in the state constitution, even though it was unenforceable.

The only group to come out against the ban was a small Confederate heritage group that seemed to be more of a one-man show than a real movement. This group claimed not to be opposed to interracial marriage but believed that an amendment to the constitution was unnecessary. The group's leader contended that the state constitution never prohibited interracial marriage, even though the part of the constitution that was removed in the vote states, "The legislature shall never pass any law to authorize or legalize any marriage between any white person and a Negro or descendant of a Negro." But few people seemed to be embracing this group's views, at least openly.

Still, the past has a way of lingering in the present. As I pondered making contact with my unknown Richardson relatives, who might be reluctant to acknowledge a familial link, it was hard not to think back to stories of denial and acceptance of issues of race that sprouted from the vote to amend the state's ban on interracial marriage. I wondered if my own encounters would take a similar tone, with my relatives noting that interracial marriage never played a role in relationships among members of the broader extended family. All the while, I was unaware that my chance for this encounter was coming sooner rather than later.

I thought I would have to track down these seemingly elusive relatives, but they chose to seek me out instead. Washington County is still rural and remote. News that an outsider, particularly one from Washington, DC, was making inquiries about the Richardsons, spread quickly through the gossip mill from community to community. As a result of this local grapevine, one connection was made through my cousin Jimmy, who still lives in the house my grandparents built. One Saturday afternoon in a grocery store parking lot, out of the blue, Jimmy was approached by an older gentleman who abruptly announced to him, "You don't know me, do you?" Taken aback by this abrupt introduction, Jimmy answered, in a cautious voice, "No, I don't believe I do know who you are."

"Well, I'm a Richardson, so I'm some kin to you," the stranger blurted out with confidence and soon took Jimmy's hand and shook it. He knew how to get Jimmy's attention, remarking that he heard Jimmy lived in the house our grandfather had built. "I knew your grandfather; even spent a bit of time with him right before he died."

The man offered to share papers and photographs, so Jimmy called immediately to tell me of his encounter. I agreed that I would visit our newfound relative on my next trip. Given the close-knit nature of Washington County, I promised to be discrete, hoping that this assurance would lead to an open and honest exchange. The man, who I will call "Tom," happened to be a third cousin of my grandfather's. Of course, Tom had heard through his network in Washington County that I was exploring some Richardson family history. A few months later on a hot summer evening I drove down a dark, remote country road to visit a heretofore unknown relative. With sweet tea and chocolate cake in hand, I found myself at the man's kitchen table poring over photographs and genealogy charts with him and his cousin, who I will call "Anna."

Anna was a petite woman, a virtual walking computer of genealogical information on the Richardson family. Armed with a steel-trap memory, ancient yellowing documents, and charts, she gave me a primer in the history of the family. "If you're going to write about the Richardsons, I want to make sure you get it right," she pointedly announced to me a number of times during the evening. Much of the history she gave me I had pieced together through my own research, though now I better understood how the branches of the family fit together. I also gained a firmer understanding of how the family had settled in the county and how my branch of the family had migrated from the original settlement at Bassett's Creek to Carson, Leroy, and Prestwick.

While my side of the family seemed uninterested in the Richardsons' role as early settlers to Washington County, the aristocratic origins of the family, with its "first family" designation, was an openly displayed source of pride to my newfound relatives. Consequently, the first document they shared with me was the will of my great-great grandfather John Richardson, which prominently featured his vast holdings of property and slaves. Anna was quick to point out to me that in the will that John Richardson provided for the care of his favorite slave:

> I further wish that my old Negro man Stephen
> at the time of the divisions of the other negroes
> choose his owner or one he would rather live with,
> to be aforementioned with the other. . . . It is my
> desire that he be treated humanely and kindly by
> those who take him.

I nodded politely as she pointed out this detail, thinking to myself while one generation owned slaves, two generations later they married their descendents.

Anna and Tom's side of the family continued to live near John Richardson's ancestral lands, but my side of the Richardson family moved to take advantage of the post-Civil War logging boom. Their families farmed and went into business in and around the town of Chatom, which in time grew into the county seat. Clearly there was a cultural difference between the families that had little to do with race. Then, something my mother's cousin Woody said to me came back: this was the part of the family that had become more linked with white society. Our meeting was probably the first time they had openly acknowledged any of their black relatives as family.

In the course of this summer evening, I began to get the

distinct impression that the invitation had been extended to me not so that I could tell the real story of the family, but so that I would tell it from a particular perspective. Anna repeatedly reminded me that she wanted me to "get it right." And she came prepared with one particular detail that she wanted to make sure I noted: my grandfather's name. She gave me a photocopy of my grandfather's draft card from World War I. His name read "James Morgan Richardson," which is different from his cemetery headstone, which reads "James Monroe Richardson." All the legal documents I had found read James Morgan, so I had always assumed that the name on the headstone was incorrect, though I could never confirm that it was in error. Of course, I thanked her for clearing up this discrepancy. Then, she asked me how the name on the headstone could have been done incorrectly. As I recounted the story behind the circumstances of Jim's funeral and burial, which had been taken care of by his surviving siblings rather than his children, the tone of the conversation began to shift from the family's role in local history to our differing perspectives on the history of the family itself.

"But as his siblings they had every legal right to determine where he was buried," Anna told me quite adamantly. She pointed out that according to Alabama law my grandparents' marriage was not legal, since it was an interracial relationship. That meant his children were not legitimate heirs and had no legal grounds to determine where he was buried or any claims to his estate.

What she said was true according to the letter of the law. Yet I expected, perhaps with hopeful innocence, that there would be some understanding of the absurdity of a law interfering with a family matter, particularly one so personal. At that moment, my face grew stern, but I contained my anger. "But they were a close family," I said with mild irritation, noting how the law was wrong and unfair. Anna had no response to that. "Plus, they had to make it clear to the relatives that they had the right to visit the cemetery whenever they wanted. Otherwise they wouldn't have buried him there."

"But it's a public cemetery," she noted with an air of derision in her voice. "No one could stop them from visiting." I was now on the verge of losing my temper. "Cemeteries were segregated in 1956." My tone was direct and authoritative. "Black people simply could not set foot in a white cemetery whenever they wanted." Pine Grove is a public cemetery today, a place I visit on each trip to Washington County. I am always made to feel welcome there by the people I sometimes meet as I stare at my grandfather's grave for insight. But when my grandfather was buried there by the Richardson side of the family, Alabama was intensely segregated. And it was segregation that landed him in a white cemetery rather than next to my grandmother.

Though it was hard for me to imagine being unaware of how segregation influenced every dimension of life in the South, even determining whether one's skin was the right color to open the gates of a cemetery, it seemed as if Anna could not comprehend how my mother's family must have felt. Although she was eager to share family history, Anna acted as if she had no understanding of the complexities that governed the everyday life of her black Richardson relatives, much less how those complexities made their lives different from the life she led. Maybe I had missed an opportunity to explain these complexities in full, but somehow I knew that she had to know what they were. We had both lived through the changes of the past forty years, but we had encountered those changes in different ways. While we could now sit together without hostility, and actually quite amiably, there was still distance between us that without question had its origin in our differing cultural perspectives. Race had constructed different perspectives for both of us, and neither of us knew how to get our minds around our differing points of view. From the way our conversation played out, I'm not even sure that Anna viewed me as a relative, who shared a common ancestry. Perhaps if she saw me differently, the very issue that divided the family in the past would not have rendered the same division in the present.

When Anna stated the rules that governed who was family, as it related to determining where my grandfather was buried, I sensed a profound gap between us. Although her statement did not directly support the divisions in the past, it pained and angered me to see her perpetuate those same divisions all these years later. Even with all the charts, wills, and documents that show our common ancestry, the fact that my side of the family is black still divides us. Paradoxically, we are part of the same family, but separate, since the family I come from existed illegitimately, according to the laws of the state of Alabama at the time. That label persists even today, years after the laws that marked us have been repealed.

Though the evening began for me with great hope for a breakthrough in understanding between the two sides of the family, it ended with a hazy feeling of loss and longing. While navigating the twists and turns of the winding road in rural Alabama, emptiness overcame me like the pitch dark I drove through. Deep down I knew that the complexities of the past could not be unraveled in one evening. But I had not expected to leave feeling that the past remained woven together so tightly and powerfully, obscuring the present, as if enveloped in the same darkness I faced on the road in front of me.

SILENCE WHEN CONFRONTING A DIFFICULT TOPIC IS OFTEN A warning sign of deeper divisions. Although no one was silent that summer evening with my newfound relations, I remained concerned by what went unspoken more than what was actually said. Neither of us confronted the true nature of the family's division, divisions whose origins are rooted in cultural differences connected to race. After some time had passed, I decided to visit my cousin Tom again to keep the lines of communication open. On my first visit I had prepared no questions to guide our conversation; I had let the evening move along in the traditional format of a Southern meet-and-greet get-together, which is exactly what it was. This time I had specific questions, as well as a bit more knowledge of the family's racial history.

Now that I knew my cousin's uncles, my grandfather's uncles, had provided testimony to prove a case of miscegenation, I wanted to know if my cousin knew about it, if it was something that was discussed in his family. He said it had not been, and neither had Edna's race been discussed in the family, nor did he feel that it mattered. I found this hard to believe, but kept pushing the question of the role of race in family relationships. He refused to discuss it, saying, "I just don't want to offend anyone."

I tried to keep the conversation going by asking what was it that bound the family together. "The woods and logging was a bond among these men," he told me. That was true, and my Uncle Smokey confirmed that his work with his white cousins was a bond for him, although the bond ended as soon as they came out of the woods. "When I saw these same men walking down a street in town, they wouldn't look me in the eye," he added.

But my cousin was not willing to confront the dual nature of family relationships that made a bond formed in the woods null and void on a public street. Nor would he discuss any issues regarding race in our family, denying that race ever played a role in our family's relationships. I stopped pushing and decided to let my cousin tell me what he thought I should know about the family. Mostly, he wanted me to be proud of the leadership role the Richardsons played in Washington County, with my great-great-grandfather being the first superintendent of edu-

cation back in 1871 and the family's role as early settlers. That
desire for education was one that Jim Richardson shared with
his extended family, since he offered to send all of his children
to college, my mother being the only one who chose to go. I
listened intently, yet what he said to me just rang hollow. It
took weeks after our meeting for me to distill what might be
behind our encounter, though my cousin had said so little that
day that there was scarce substance. When I took my confusion
and thought of it in the context of the traditional mindset of
the South, things began to make some sense, although they still
seemed a bit out of focus.

My cousin was affable and charming, yet his Southern manners would not allow him to lift the veneer of politeness to have
an open discussion about the role race has played in our family
history. Like most white men of his generation, my cousin had
simply never had to engage in a conversation about race. Race
is traditionally a topic one does not bring up in polite company
in the South. If he had engaged with me in an open discussion
of race, he feared that he would offend me in some way, leading
to his response, "I just don't want to offend anyone."

The entry point for most conversations about race begins
with what two people have in common, which I thought would
be easy given that my cousin and I share an ancestral bond, one
he is eager to explain to me in great detail. But there is another
leap of consciousness that must be made for the two parties to
engage in an honest exchange. The next leap is moving beyond
race and rendering it invisible, and the two of us just could not

get there. For my cousin, having me come to his home as a welcome guest, and his open acknowledgment of me as a relative, was his leap into the present; he simply lacked the level of awareness of racial issues to acknowledge that the family's racial divisions ever had any relevance to his life, either then or now. Nor did I think he wanted to make the leap it would take to enhance his awareness.

Open conversations and discussions happen when people decide to speak directly and honestly. Unfortunately, my cousin remains stuck in his way of thinking about race as a taboo subject for an open discussion, making it difficult to move our encounter into a real conversation. And I felt stymied as well, bogged down by my desire to be honest yet not to cross a line dictated by the rules of polite discourse in the South. What seemed like a beginning became mired in an outdated way of thinking about race that even forty years of changes in race relations cannot break down without difficulty. The taboo was far too ingrained in both of us to overcome the resistance to speak candidly.

AFTER MY TALKS WITH MY COUSINS WHO REMAINED IN WASH-ington County, I began to cast my net wider for different perspectives in the family, with the hope that I would not be met with the same fear and silence again. Since inquiries about race and family would undoubtedly be faced with silence in a phone inquiry from a relative stranger, I chose a different tactic: an

old-fashioned letter. In my letters, I explained my general purpose and exactly how I was related to the recipient. Then, I
introduced the topic and was clear about what I was looking for:
differing perspectives on the issue of race from the black and
white sides of the Richardson family. I continued, "Through
interviews with family members from both sides of the racial
divide, I hope to create a cultural biography, placing the Richardson family in the broadest possible social and historical context. At the same time, I hope that together we can reconstruct
a concept of family that is no longer based on race."

A few days after sending the letters, I would follow up with
a phone call to set up a time to meet. I soon discovered that
communicating in writing was surprisingly less effective than
a phone call: my invitation was routinely met with refusal. And
when someone would agree to sit and talk, a phone call would
come several days later saying he or she had changed his or her
mind. Usually I would hear, "I'm not sure I want what I have to
say to go into a book" or "I'm a private person." Others would
just say that they didn't think what they would have to say
would be of any help. When I explained that our conversation
would not be a game of "gotcha," but my personal attempt to
broaden the conversation about race, the response was silence,
followed by some hedging and then another outright refusal.
Then, I would try again by saying, "Let's meet and just make
the family connection," hoping that the personal connection we
might make would foster a genuine conversation and build trust
to talk honestly about the role race has played in relationships

in the Richardson family. But even using this initiative, I was always refused outright.

After a series of letters, phone calls, and refusals to talk, I was overcome with moments of confusion and sadness, followed by anger and frustration. Although all my encounters were characterized by genteel Southern manners and no one was rude, I thought it was better to be honest about what I wanted to talk about rather than to present myself simply as a long-lost relative reaching out to a family member. Our family connection was much more complex and tied in with events in the recent and distant past having to do with race and how it divided one family. The sadness came because the people I wanted to talk with about race, though they were strangers, shared a blood tie. Perhaps it was naiveté combined with arrogance that made me think our common kinship would make a difference in both sides reaching across a racial divide to the other.

Although the laws that barred my grandparents' marriage are gone, the social residue of those same laws continues to taint the way family members in an interracial family view race and racial identity. What remains of this social system is who gets treated as family and who does not. The failed encounters with some of my white relatives are the remnants of a line that was legally established to divide the family, and that line still exists in the minds of many of its members. That line still divides us, leaving us in separate worlds and separate families, seemingly lost to each other forever.

III

TRANSCENDENCE

My old man's a white old man
And my old mother's black.
If ever I cursed my white old man
I take my curses back.

—LANGSTON HUGHES, "CROSS"

9.

Transcending Ambiguity

"YOU'RE FROM A MULTIRACIAL BACKGROUND, AREN'T YOU?" commented a relative stranger I spoke with one evening at a cocktail party several years ago. Not offended but taken somewhat aback by this proclamation, I maintained my composure. "No, I'm black," I replied, pointing out that a multiracial identity did not exist when I grew up in Mississippi in the 1960s. It simply wasn't part of the cultural landscape; plus being black is a part of who I am, how I view the world, and the culture that shaped me. We went on to have a friendly and honest exchange about how racial identity is much more fluid and complex today than it was when we both grew up. Today, someone might look at me and see a man from a multiracial background; forty years ago, you would see only a black man. The more we talked the more I realized that my cocktail party companion genuinely understood how our much more diverse contemporary culture leads people to embrace not just one, but sometimes several cultural identities. At the same time, she understood how the culture of my native Mississippi during the civil rights era formed my personal identity and allowed me to see myself less fluidly than a younger person from a similar background.

Now, after closely examining the multiracial origins of my family, I would still answer my companion's question the same way. My personal identity has not shifted with this deeper knowledge of my family's racial history; what's bred is in the bone. Identity and a strong sense of self form the foundation of the fundamental character of my family's heritage, and I am grateful for being part of a legacy that includes transcending racial ambiguity for the sake of family survival. At the same time, I would not deny that I am a black person from a multiracial background, because that multiracial background shaped my black identity as much as my maternal relatives' decision to embrace blackness. The decision of Jim Richardson and Edna Howell to live as black people and the values they passed on helped me to see white people not as the unknowable, superior other but as equals. The ambiguities in my personal history and outward appearance are as much a part of who I am as is the surety of my personal identity.

Traditionally, American culture does not embrace ambiguity, particularly when it comes to race. For much of our existence as a society Americans have held on to an all or nothing sensibility when it comes to identity, with little tolerance for those who may be caught between cultures or racial identities. Yet, it seems as if cracks are beginning to form in our sensibilities, allowing many of us to accept racial ambiguity rather than trying to categorize people firmly into one racial or ethnic group.

But this will happen slowly, given that we've spent hundreds of years eliminating complexities from the equation of racial and cultural identity.

The desire to categorize people into specific racial or ethnic groups is one cultural remnant of slavery that cuts across American society, but is one that has cut the deepest among those with African ancestry. This cultural relic shaped the identity of the children of Jim and Edna Richardson and subsequently my own, leading all of us to adopt a solid black identity. Many in the family have outward characteristics that reveal a multiracial background and others appear to be white, yet the family's tilt toward blackness has been unwavering. For those in the family who appear to be white, racial passing was rejected out of hand; sublimating one identity to fabricate another was considered to be a path to the worst psychic traps imaginable. In addition, cultural and legal forces made living a racially ambiguous life nearly impossible as well as undesirable, given the color consciousness of American society and years of pervasive Negrophobia.

For almost 200 years, American racial identity has been governed by a series of legal acrobatics that determined which racial category fit an individual. The best example of the cultural and legal forces governing personal racial identity can be found in the historic case of Homer Plessy. In 1890, Homer Plessy challenged a Louisiana law ordering the strict segregation of railroad cars into white and black sections. Plessy looked white, in spite of being one-eighth black, and he argued that he should have the same rights as white citizens because of his skin color. If the purpose of a Jim Crow car was to separate people by color, then with his European features and skin color he fit the bill for riding in the white car. However, the United States Supreme Court failed to see the absurdity of Jim Crow laws in particular and the racial caste system in general. Moreover, the Supreme Court noted that there were "physical differences" between whites and blacks, as well as different "racial instincts" that would require the races to be separated. In spite of a dissent by Justice John Marshall Harlan, declaring that the United States Constitution was "color blind," the Supreme Court ruled against Homer Plessy 7 to 1. The end result was that its decision engrained the "separate but equal" doctrine into American life for more than half a century.

While the *Brown v. Board of Education* decision wiped away the separate but equal doctrine established by *Plessy v. Ferguson*, the mindset that led to the Plessy decision lingered in American culture and perpetuated a racial caste system based in the one-drop rule that served as the core of Plessy's argument. That same mindset came into play in Alabama's miscegenation laws. The end result determined, without a shadow of a doubt, the Richardson family's choice to live as black people and not look back.

Americans have lived with the cultural sensibilities of a racial caste system for more than a century, but now there are signs that the remains of this relic are beginning to disintegrate. With an increased multiracial population, American culture may be forced to accept the ambiguity that *Plessy v. Ferguson* sought to eliminate. Advertising routinely features people with racially indeterminate features as well as interracial couples. Prominent scholars such as Evelyn Hammond, a professor of the history of science and Afro-American studies at Harvard, are proclaiming race a human contrivance, a "concept we invented to categorize the perceived biological, social and cultural differences between human groups."

But is race an entirely human contrivance? There are biological reasons for why people have dark or light skin, certain facial features, or hair texture, and some components of the physical features of different ethnic groups can be seen in the faces of people we meet every day. All of these characteristics are encoded in something we cannot see: our genes. And sometimes characteristics are buried in our genes that are not reflected in our outward appearance. DNA testing reveals that 5 percent of white Americans have some recent West African ancestry, and many black Americans have white bloodlines, after years of various schemes of racial demarcations based on skin tone and one-drop rules. In spite of this scientific evidence as well as our society's claim to have multicultural aspirations, we continue to think in terms of outmoded racial categories, based on the one-drop rule, as well as stereotypes influenced by ideological or cultural influences, such as the "tragic mulatto" whose mixed-race background leads to a life of confusion and suffering. In our focus on cultural differences we've come to focus more on what divides us rather than what unifies us.

Racial categories have never followed any clear logic for biracial and multiracial individuals, either today or in the recent

American past. More likely than not, race will always matter in American culture. The bigger issue is how we can make race matter less. Perhaps the best way is by embracing racial ambiguity while at the same time transcending it by not allowing it to be so tightly defined and classified.

The scientific world of genetic ancestry tests may provide a means of making these categories and cultural myths meaningless. That is, if we can discard the old eugenics-based system of calculating percentages of blood and racial hierarchies. To do that, we must look at genetic ancestry tests as a means of thinking differently about race and identity. But that requires separating science from cultural lore and stereotypes, which is equally difficult.

In 2005, scientists reported the discovery of a genetic mutation that led to the first appearance of white skin in humans. When I read about it, I wondered how it is that a minor mutation—just one letter of DNA code out of 3.1 billion letters in the human genome—became so highly prized that it has served to divide people for generations, including members of my own family. Discovery of this mutation, combined with recent findings that all people are more than 99.9 percent genetically identical, reinforced my belief that race as we know it today is almost entirely a social demarcation. While DNA reveals that there is a biological component to the physical characteristics of race, the biological component is entirely separate from the social and cultural components that most Americans identify with race and that have no basis in biology. The idea of racial instincts that played a role in establishing the principle of separate but equal in this country is still ingrained in our cultural sensibilities.

To answer some of the questions I had about my personal concept of race, I decided to have my DNA analyzed to determine as precisely as I could where the various components of my ancestry came from. Sending away a DNA sample taken from a cheek swab and a small blood sample seemed to be the right thing to do, particularly as I began to think about what I had in common with my white Richardson relatives and what it was that made me different from them. When my results arrived, I immediately turned to the "Ancestry by DNA" results from DNA Print Genomics to look at the percentages: 60% West African, 32% European, 6% East Asian, and 2% Native American. There were no real surprises, except for the East Asian percentage. With so much discussion of my grandmother's Native American ancestry, I thought that more of my DNA would be American Indian rather than from East Asia. Generally there is low Native American ancestry among non-Hispanic Americans, both African Americans and those of European descent, but the test made it clear that I carried those genes. But as I studied the information that accompanied the results, I realized that if I went back five generations, there would be 32 great-grandparents who would be passing on their genetic material to me. The equivalent of one of those great-grandparents had some genetic link to Asia. They might be Native American or they might not.

There was also a chart with a triangle graph and a circle plot, but I confess that even after reading the accompanying materials I was perplexed. A bright red dot within the circle plot on the side of my Asian ancestry marker, indicating the maximum likelihood of the percentage, simply added to my confusion. From looking at the chart, I could not discern whether the likelihood was high or low. Although I had once taken a college course in genetics and found the science of DNA and genetics fascinating, my knowledge was thin and superficial, much like the grade in genetics I once received. Equipped with the knowledge and sensibilities of the average layperson, I realized that identity as prescribed by numbers had been imprinted on my sensibilities. This closely held cultural concept led me to focus primarily on the estimated percentages of genetic ancestry rather than the more sophisticated interpretative scientific approach the graph represented.

While moving between the graph and the estimated percentages, and in the end concentrating more on the percentages, I began to see how the one-drop rule influences much of our thinking about race. Whether we are facing a logical scientific test based in complex statistical analysis of results or looking at a person's outward appearance and guessing whether he or she is half-, one-quarter, or one-eighth black, those sensibilities come into play like a hidden instinct seared into our minds. In spite of this realization, I wondered where this East Asian ancestry had come from. Asian ancestry did not fit into the legal definition of race in Alabama or anywhere in the American South, where everything was black and white. This piece of information also did not fit in with my notions of myself or the notions of the ancestry of either of my grandparents and fogged up any attempt to do my own interpretation of the data.

The answers to my questions must lie in the mysterious triangle graph, I thought. And I wanted to erase a culturally conditioned concept of race and identity, with its proscribed percentages and generational calculations of racial blood, and move to a more nuanced and scientific perspective on the topic. I turned to Pennsylvania State University anthropologist Mark Shriver to analyze this chart and help me see my DNA the way a scientist, rather than a layperson like me tainted by cultural misperceptions, would look at them.

As I waited outside his office, I examined a myriad of Polaroid photographs that covered the walls. Each face was accompanied by an ethnic designation that matched the personal identity of the face in the photograph. Beside the photograph was a paper flap to lift up to see exactly what a DNA sample revealed about the person's ethnic background. As I went through photograph after photograph, I was surprised that few of the ethnic identities of the people actually matched up precisely with their DNA. With a few exceptions, most of the people had some mixture of DNA from at least two groups; many were like me and had genetic ancestry from Europe, East Asia, West Africa, and Native American groups. There were blond people with African and Asian ancestry, while there were several dark-skinned people whose results revealed that more than half of their DNA came from Europe.

When I sat down with Mark Shriver, I commented that

the mosaic of photographs showed that DNA does show that we're not always who we thought we were. "There is often a gap between your personal identity and where your genetic material actually came from. There's a lot you can tell by looking at someone, and then again there is a lot that you can't tell by looking," Shriver said. He mentioned that his own genetic ancestry analysis—13% West African and 5% Native American—contrasts with his outward appearance. And I admit I would have never guessed that he had that much African ancestry by just looking at him.

As we examined my results, Shriver told me exactly what I had thought: just looking at the numbers is not the best way to interpret DNA results. The numbers often lead people to think, not in terms of genetics, but more along the lines of the old one-drop rule. For my particular results, he explained, the West African and European percentages were fairly accurate. Judging from the graph, those percentages would mean that slightly more than half of my DNA is from West Africa with the next substantial genetic component of my background coming from Europe. The 2% Native American and 6% East Asian ancestry percentages, however, are misleading. The way the circles on the triangle plot graph fall are consistent with someone who probably has more Native American ancestry than the percentages indicate. The circles move in both directions around the axis of the graph for East Asian and Native American, offering compelling evidence that I have ancestry from both groups, or from either. However, taking into account my family history, which includes Native American ancestry, it is more likely that for this particular component of my background I have more Native American than East Asian ancestry. "Keep in mind that both of these two groups are the most closely related evolutionarily. Both groups had their origin in central Asia. However, we do see different levels of central East Asian ancestry in some Europeans, so some of that East Asian ancestry could also have come from your grandfather."

That would mean that my grandfather was not 100 percent European, an idea I know he would have liked and that I find interesting, since it compounds the folly of the miscegenation laws that made my grandparents' marriage illegal. Still, Shriver suggested that if I wanted to delve more into the question of my Native American ancestry, I might want to consider a mitochondrial DNA test. It might help confirm the stories I had heard of my grandmother's Creek Indian ancestors. It's interesting, but I wonder if I already have compelling evidence of this ancestry: Native American characteristics show up in my children's faces, particularly in my son Aidan. That's good enough for me. But Shriver reminded me that it's hard to isolate particular facial features and directly tie them to a piece of your genetic ancestry: "We only look at facial features for generalities, not specifics."

Knowing more about my grandmother's personal history through DNA is compelling, since there are so few other parts of her life that I can piece together. One part of me wanted to unravel the Native American component of my family so that I could feel some closeness to her and her ancestors. But I do know that her Native American ancestry wasn't the core of her identity. Exploring this Native American link was more about me than it was about her, which revealed how the results of a DNA test could lead someone to obsess about various aspects of his or her racial background. The obsession might increase if that person did not know as much about his or her background as I know about mine or how the people who carried those traits thought of themselves. I asked Shriver if DNA analysis makes us focus more on race, as I did with my Native American and East Asian ancestry, or does it deconstruct current cultural notions of race, as revealed in the wall of photographs?

"We don't think in terms of 'race' in biology," Shriver noted. "And there are lots of problems in talking about humans in terms of race, since race has been a point of debate throughout the twentieth century and even before." The father of physical anthropology, Johann Friedrich Blumenbach in his 1776 book *On the Natural Varieties of Mankind* wrote about how groups intersected and used geographical labels, not racial labels, to distinguish people. Blumenbach was the least racist of Enlightenment writers on human diversity, yet in spite of his commitment to human unity, over the centuries his geographical scheme was used to place people in racial groups based on physical beauty and intellectual worth, leading us to the early twentieth-century concept of race that still pervades much of our thinking today.

"I'm actually uncomfortable generally using the term 'race' unless I'm talking about it when it's culture and biology together," Shriver added. Race is a matter of prioritizing for value rather than nonexistence. What I believe we should all do is focus less on the outward characteristics that make up what we think of as race and take into account the cultural experience more. That makes us think less about the old idea of race and more about culture and cultural experiences." Shriver reminded me that race implies—at least as we talk about it in the United States— that we are one race and not the other and can be used as a marker to separate people from each other. DNA analysis turns that notion upside down and reveals the archaic nature of the traditional American concept of race based in standards of purity and superiority. "The philosophy for looking at human life and its value in American culture was shaped around slavery," Shriver noted. "It was to the advantage of those in power to have a subhuman group of people."

It is that concept of race rooted in slavery that plays into how the black and white sides of the Richardson family look at each other. Our separation is uniquely American, Shriver pointed out to me, because the United States is the only country where race is established by family. "Because of the one-drop rule, this familial sense of race became culturally embedded. And because we are not that far removed from the Jim Crow era, it's going to be difficult to remove that embedded concept from our cultural consciousness." Our language and discussions about race and racial issues continue to be littered with references to the scientific construct of race that came out of the era of eugenics rather than seeing race as a combination of culture and biology. The myth of the genetic inferiority of people of

African descent clouds the minds of some members of the white Richardsons and is one of the issues that created the racial divide in the family that lingers today. Plus, the idea that race is something you inherit has only served to keep the division in place. Still, I wonder if giving people the opportunity to look at their ancestral background through DNA analysis will change this concept of race or just get them to think in percentages, much as I had when I first got my DNA analysis.

Then, Mark Shriver showed me that the human genome has 100,000 parts. The parts that make us physically different—skin color, hair, certain facial features—are only a small part of it. He firmly believes that DNA helps to break down the categorical way people think about race. When you see how small the differences are in relation to the entire human genome, it's hard to make a fuss about the small things that make us different. "You can diffuse traditional thinking about race by making people see these differences as natural and teaching them that the differences are just part of the variety of life. That's the trajectory we are on regardless. How quickly we can get there depends on how good of a job we do in educating people to this new way of thinking." People must begin to think in terms of the science and to disassociate issues of ethnicity and cultural heritage from the biogeographic ancestry they might discover from a DNA ancestry test. But Shriver also reminded me of something: we really cannot dismiss race categorically. To dismiss race is to ignore the beauty of the variability and variety that exists among people. "What's going to happen to us genetically in the future is similar to what happened to us in the past. We'll be able to know where we are in our place in the universe by studying these aspects of human variation." Acknowledging the beauty of racial differences makes them less meaningful as dividing lines and more meaningful as windows to our pasts and futures, since DNA reveals that we are all more closely linked than we appear to be.

When I saw the large number of common genetic markers shared by humans versus the ones unique to specific racial groups, something began to click. Combined with the mosaic of faces and all of the various ethnicities posted up on the wall, I began to separate ethnicity from actual biological ancestry. This scientific look at race made me think differently about what was race and moreover what the idea of race meant to me. At one time, race in my mind was a dividing line, a social construct, a source of tension and pain. It was both cultural and political, woven into my consciousness as much as it is into the very fabric of American society. Yet the line of race no longer pervades my sensibilities now that I know where that same line that traditionally has divided people joins together with a common humanity we all share. For me, race has evolved into a social artifice, something that is physically and scientifically real but largely constructed by our broader society through what we project on a person's physical characteristics.

Race has also moved from being something concrete and indissoluble to a construct that is more amorphous and fluid. Nothing has shifted in how I perceive and present my personal identity. Seeing it plotted out on a graph has just made it mean less to me now that I can see more of what I share with others who do not necessarily look the same. I can no longer look at people and regularly ask myself, "what are they?" or discern their backgrounds by a cursory glance. I've learned that there is more to ethnic and racial backgrounds than meets the eye. On the occasions when I do shift into my old way of thinking, I now tell myself, "why does it matter what they are?" and move on.

For me, as for most Americans of my generation, race will never be completely fluid and meaningless. It would be practically impossible to strip away years of social conditioning as well as to peel away the veneer of race that covers so many social interactions in this country. But race is more fluid, less meaningful, than it once was, particularly among younger people. And as this new way of thinking begins to take hold, ideas about race flow differently from my mind, much differently than they did as a teenager when I lay in my bed and repressed my thoughts about my grandfather being a white man by thinking it had nothing to do with me. Now I finally see how undeniably he is a part of me and how we are alike in more ways than we are different.

Race remains the topic that many dare not discuss and ignore and disregard rather than acknowledge as a cultural reality that cannot and will not disappear. It's imprinted in our DNA, shows up in our faces, the shape of our eyes and noses, and, yes, in the color of our skin. There is no denying that it is an inescapable framework for shaping our perceptions of others, as evidenced by the fact that race remains an essential component of individual identity and government statistics. Consequently, even if you have a new way of thinking about race, one that views race as a reality yet an insignificant reality, the broader culture has ways to keep you bogged down in your old way of thinking. So, issues of race, color, and ethnicity will not disappear. Nor should we think idealistically that new scientific and social constructs can make the idea of race go away. But a new way of thinking can make race a less significant factor in the way we live our lives. It can also change the way we treat other people. Somehow, in spite of living in a hyper-racialized society, Jim and Edna Richardson moved beyond race almost 100 years ago. I'll never really know exactly how they moved beyond race, but I do know that what they did played a role in changing the way I think today.

Once the scientific insignificance of race comes into focus, cultural myth will play a weaker role in one's perceptions. It is those cultural myths that have divided the Richardsons and shaped their perceptions of each other. Seeing and understanding how our entire DNA is intertwined helps make the cultural concept of race seem arbitrary and artificial. And when race becomes less important, it makes us focus on what is shared and universal, which is what Jim and Edna Richardson did long ago. From Jim and Edna I inherited my genes and a rich family history, not my race. For me, recognizing that has made all the difference.

People living in formal societies, lacking the historical imagination, can imagine for themselves only a timeless existence.

—ALLEN TATE, *THE FATHERS*

10.

Moving Beyond the Myth

FAMILY IS CENTRAL TO SOUTHERN SOCIAL LIFE AND IS braided together into a complex knot along with history and myth. This interconnected legacy has a way of distorting history, leading to the creation of even stronger myths that seem impossible to destroy. And so it is with the Richardsons. Race, while seemingly a black and white issue in the family, is far more complex when the idea of cross-racial kinship comes into the picture.

Although myth can sometimes overpower reality, there is no denying that genetic material from four continents came together to create the person that I see in the mirror each morning. And that is reinforced when I glance at Jim Richardson's picture alongside the portraits of all my other relatives on my living room wall. Some of his genes are revealed in my face and physical features, while some remain artfully masked by characteristics from other relatives. These pictures, combined with what I know from my DNA, tell the story of who I am. Other characteristics are mirrored in the faces of the relatives I've encountered in my travels around south Alabama. Over time, this scientifically and culturally heightened awareness of my background has moved my concept of race into a broader framework, outside of the realm of myth. Perhaps that is because I have come to recognize the dangers posed by embracing the myths, which fix your mindset in the past rather than the future.

Examining race through the science of DNA analysis may have reconfigured my thinking and crushed some personal myths, but conversations across the racial divide of my family did not reveal a similar movement. For the family members I have encountered over the years, cultural differences are much more strongly rooted and tangible than the science of DNA. The major divide in the family arises from a sense of family rooted in the concept of race based in the laws and customs of the old South. My idea of moving beyond race or thinking about race differently undercuts two pillars of thought once deemed timeless and immovable: the one-drop rule and prohibitions against intermarriage. And that is the origin of the cultural gap I have encountered in many of my conversations.

Cultural gaps and misperceptions often divide perspectives on race and not just in families with multiracial origins like the Richardsons. The black Richardsons, while secure in their personal identities, have always recognized the familial link to their white relatives. They never felt compelled to hide it. Concealing their mixed blood was impossible since that link was revealed publicly in where they lived and how they lived their lives, even if it was not revealed in their skin color and features. Conversely, many white members of the family were conditioned not to recognize their black relatives as family, in spite of familial ties that would be blatantly obvious. Even those who did recognize them as family often fell into traditional black-white social roles that called for blacks to behave deferentially to whites. Though this seems archaic today, we must recognize that at the time racial differences were viewed to be so profound that acknowledging family ties with black relatives would undermine your social standing as well as your view of the world. In the United States, particularly in the South, kinship across racial lines was deemed abnormal, outside of acceptable racial and social etiquette, something that was better left undisclosed and unspoken. It is these two ways of thinking that characterized many of my cross-racial conversations with Richardson family members.

While cultural differences and mindsets play a role in many of the conversations I have had with white family members, there are generational differences as well. In the older generation, some of the white Richardsons have genuinely made a shift that allows both sides to talk and recognize a once denied or ignored family tie. Still, remnants of the past remain at work, leading them to deny that race played any role in family relationships in spite of evidence to the contrary. Although cross-racial conversations are now acceptable, a true shift in thinking has not yet taken place and probably never will. Among my generation, the two sides acknowledge each other yet rarely bridge the divide that separates us. Though we were shaped by integration and the civil rights movement, true social integration eluded us. Denial does not characterize our conversations, but there is discomfort wedged between our hollow silences. We don't know how to have an honest discussion about race. But the generation that follows us acknowledges the familial links between black and white without shame, guilt, or embarrassment. This generation's lives are defined by fluidity and connectedness to people, making cross-cultural conversations less daunting.

It has taken more than 100 years and three generations for the attitudes to evolve to this point. Yet, in spite of all these cultural and generational differences, whenever I have had an encounter with a white family member, I have rarely felt any ill will. Some have even embraced me as family. Although this is progress, we are still strangers when we meet. When conversation comes to the issue of race, the response varies from utter silence to perplexed confusion to genuine attempts at understanding and reconciliation. Not everyone knows what to say; consequently, rather than say something wrong, oftentimes nothing is said at all. Those who can engage in the conversation have somehow forged a broader frame of reference for racial issues in their life experiences, whether consciously or unconsciously.

What these conversations have made me realize is that our cultural vocabulary related to race and racial issues remains painfully limited, and that includes my own. In each conversation, our frame of reference remains fixed in another era when race was not a topic for open discussion. Consequently, the debates and verbal parrying on racial issues rarely update the issues of the past and bring them into the present, leaving the various talks littered with missed opportunities to update the cultural conversation about the role of race in our family.

And in many ways, the Richardson family reflects the broader perspective on race in American society: the American framework for thinking about race is frozen, inflexible, and seemingly impenetrable. Yet these conversations demonstrate that if we remain frozen in our old paradigm of superior versus inferior groups and cultural myths, we can never move toward a future where we see our differences as part of the variety of life. Given the changes that are afoot in American society, our culture's future lies in moving beyond our racial myths rather than perpetuating them.

All of this means that Americans need a whole new vocabulary for talking about race, one that shows an awareness of current cultural issues, yet does not ignore what happened in the past but tries to learn from it. By examining the racial past of my own family, I have struggled to not focus on past wrongs, but to learn from them in a way that my own thinking about race can move forward, not backward. In my own effort to move forward, the life of Jim and Edna Richardson has become my guide. There is no doubt in my mind that race played a role in the way they lived their life. Given the time and place in which their relationship came about, it had to. Denying that race defined their daily existence would only shroud their life in even more myth. In fact, ideas of race developed in the eighteenth century shaped the identity of the family as well as the roles each of them had to take in their marriage. Yet in spite of the confines of their times and circumstances, they broke a set of cultural conventions that continues to reverberate through the family. Without Jim and Edna's example, I could never have come to the realization that even in the face of the cultural differences known as race, what people have in common is more important than those differences.

What we have in common is our humanity. After focusing on what we have in common, it is easier to talk about the differences that have historically divided the races. Racial progress has been made, but the legacy of segregation, miscegenation laws, and other issues continue to hinder us from recognizing our common humanity. We are only a generation removed from the days of Jim Crow and laws against interracial marriage, and both of these issues have left gaping wounds in our hearts and minds, wounds that have not yet healed. When we talk about race, we must acknowledge the wounds but not allow them to overpower the conversation to the point that our talk focuses entirely on past hurts and wrongs. It is this inability to acknowledge these wounds that has been the largest impediment to conversations across the racial divide of the Richardson family.

This means that when we talk about race, we must not brandish the term "racist" at every turn and give thoughtful examination to both white and black perspectives, recognizing the validity in both. At the same time, talks about racial issues must acknowledge that race often divides people, and will continue to divide people, and racism exists within segments of American society, even within families. As I have heard many times, racism is one of America's original sins. Although racism is a sin that must be acknowledged and confessed, we must not wallow in it unnecessarily.

The philosopher Kwame Anthony Appiah notes, "the points of entry to cross cultural conversations are things that are shared by those who are in the conversation. They do not need to be universal; all they need to be is what these particular people have in common." Appiah reminds us that real discussions happen when people decide to find new ways of thinking, feeling, and acting.

We need a path toward new ways of thinking, feeling, and acting about race, and there is much work to be done. Thinking and feeling differently about race is one step. The real test involves acting differently and building a new vocabulary for talking about race in how we act as well as how we communicate with each other.

I decided to begin to build my own vocabulary about race

by returning to talk with those white relatives who would engage with me on any level, even those who rebuffed my appeals to talk about race. Even though I thought conversations were at a stalemate, I wanted to bury some of the anger and resentment that I felt in our previous conversations because my relatives could not speak about the role of race in the family in what I thought was an honest way. With springtime in south Alabama blooming in all of its glory and promise of renewal, I returned to Washington County to begin again.

In my previous conversations, my questions were sharply aimed at the role race played in family relationships and in how race influenced the ways Jim Richardson and his family were viewed in the broader community. Perhaps my approach was too direct, I thought. Plain speaking is a Southern value that is appreciated except when the issue is as fraught as race. This time, I decided to gently sit and talk, "just visit" as people in the South say, and see what came up in the conversation.

But little in the South is forgotten. When I began exploring the life of Jim and Edna Richardson, a feature article about my efforts to reconstruct the life of my grandparents appeared in the Mobile *Press Register*. While the article was positive, I was direct in my reason for seeking to understand my family's interracial origins. I said: "Family in the South is bound by race rather than blood. It's time for it to be all by blood." The reporter and I had also visited the all-white cemetery where my grandfather is buried separately from my grandmother. I wanted readers to recognize the depth of the family's racial divide, which exists in spite of how cordially I have been treated on my visits:

> He gets out of his car and walks through the graves, many headstones with the name "Richardson" on them. Some have small Confederate flags stuck into the ground.
>
> He arrives at his grandfather's plot. "By burying him here," he says, looking at the headstone, "it's almost as if, in a way, they were trying to wrest him away from our family.
>
> "But now I'm taking him back."

Looking back, my approach was far too direct for Southern tastes. The language I used did not point to what I shared with the people I wanted to talk with. By mentioning family I thought I had brought up something that was universal and could be shared. However, talking about tying families together by blood rather than separating them by race, along with pointing out the Confederate flags in the cemetery, pressed all the buttons that make some white Southerners squirm. On this trip, I resolved not to bring up the issue of race. I wanted to see if this shift in the conversation would make a difference.

As it turned out, the decision was not mine to make; race sat firmly planted in the front of everyone's consciousness. Just a few weeks prior to my arrival in Alabama, Senator Barack

Obama had given a speech on the role of race in American society, and people, both black and white, were talking about it in churches and in letters to the Mobile *Press Register*. No matter the political persuasion of the people I encountered, Senator Obama's speech came up without my introducing it into the discussion, much less the issues of race and cultural identity. They agreed with Senator Obama that race is an issue we as a country cannot afford to ignore, but many still remained taciturn about the subject in the Richardson family. While the idea of broader racial reconciliation was on their lips, it was still hard to talk about race when it was close to home. Yet for others, the speech opened a door to an honest dialogue, with the same direct approach I took to present the topic originally.

"What is it you really want to know?" one relative asked, revealing a genuine desire to talk along with a tinge of resentment revealed in his next question. "Are you just here trying to stir up trouble?" The answer was "no," but he remained reluctant to engage, since he remarked that he still had a copy of the newspaper article written about me from eighteen months ago. "What you said there still sticks in my craw," he confessed. Apparently, the sense of a troublemaking outsider from Washington, DC, had begun to swirl around these lonely country roads my grandfather once wandered and arguably controlled. No one had ever publicly questioned, in print no less, why Jim Richardson had to be buried in a white cemetery. No one had talked openly about Jim Richardson's life in a black community where he was accepted and embraced or asked why he and his children blurred the racial lines in the eyes of many. And in any case, talking about race in this part of the country is synonymous with stirring up trouble. But people were also clamming up because they thought I had come to tear down something they were clutching on to for dear life: their local history and myth.

"Son, your granddaddy wasn't a saint," one man told me. I could not have agreed more. I told him how I knew my grandfather had killed at least one man, and probably a few others I didn't know about. For more than thirty years, he had eluded the law by illegally selling whiskey, even right out of the back of his truck in broad daylight on the streets of nearby towns according to what people had told me. And he made a fortune from this whiskey and used his fortune to educate his children and to amass land holdings that were the envy of many. While this didn't seem bad to me, it did to others. On top of it all, he had been able to carry on all of this activity by currying influence with powerful people in exchange for keeping their secrets. I had no desire to recreate or repackage his image, and once I made that clear, our conversation progressed into something real and substantial. And I learned something about my grandfather that I had never really known.

What I didn't know was that Jim Richardson lived on in Washington County as folk hero and legend who was demonized by some and admired by others. Some of the resistance to talking with me arose because there was fear that I was taking over local culture for my own purposes: trying to create a new myth of my own. The silence I faced originated in the idea that my travels to Washington County were for the express purpose of either burnishing Jim's reputation or dismantling it in their eyes by proclaiming him a race relations pioneer. They wanted no part of creating what they saw as a new myth that would replace the existing folklore. Whether he was folk hero, legend, or scoundrel remained at the center of the debate among white people around here, especially those related to him, whether that relation was close or distant. How he was to be remembered was up to them, not me. Jim Richardson's place was not in the written history of Washington County. Instead, his place was in local legends; local people, including his relatives, had no interest in arriving at truths about him or his character that may lie buried beneath the myth.

By not asking people to think or talk about race, I stumbled upon a division that had its origins in local culture and not just in race. White relatives, and some whites outside the family, genuinely feared one of their own being recast as a man who defied the boundaries of race. The idea of a local figure becoming a symbol of racial transcendence and equality generated indifference as well as fear, particularly when the subject was a man as flawed as Jim Richardson. For them, this would mean that a new dimension would be added to the local folk culture, a dimension that elicited as many mixed feelings as it did resentment, since this was a man who violated every rule of local society. As the preacher remarked at his funeral, "Jim Richardson lived his life the way he saw fit rather than the way you may have wanted him to live it." Some still were uncomfortable that Jim Richardson and his family had lived outside of the proscribed racial boundaries and that the entire family thumbed their nose at customs that were designed to separate the races.

While for his white relatives Jim Richardson stood as a symbol of folk culture, black relatives and black people saw him through a completely different lens. Those who knew Jim Richardson valued him more for his fairness and ability to see beyond race in all of his interactions than they did for his contribution to local color and legend. In spite of his flaws, he was the type of white man they all admired. During the hard times, the people of Prestwick could depend on Jim Richardson when other whites would simply ignore them or treat them with condescension or disdain. Some even said, "We needed more Jim Richardsons around here."

Although Jim Richardson was a Washington County folk hero and legend, I reminded the people I encountered that first and foremost this man was my grandfather. I wanted to understand him in his totality, both the good and bad. His blood courses through my veins and he reveals himself in my DNA. If he had not been able to move beyond race, he would have been a nameless, faceless, unknowable white man who through happenstance and concubinage came to be my grandfather.

Instead, he made a home with my grandmother, loved her, and provided for his family. Jim lives on in his three surviving children, all of them nearing the ends of their lives and all of them still professing their undying love and affection for him. To them, his ability to cross and bridge racial boundaries was as much a part of who he was as was his more scandalous behavior. He was nobody's saint, but he wasn't all sinner. In the eyes of many, his redemption as a man derived from his genuine desire to be seen not as a white man, but just as a man. He continued to gain privileges from his whiteness, yet he rose above his status and race to be seen as just a man. To me, that sounds like transcendence.

Few of us know how to genuinely transcend racial differences. Jim Richardson succeeded at moving above racial distinctions with his family and the people of Prestwick, yet he could not move beyond race outside of those confines. The social forces were too strong. But those social forces have long since crumbled, though by our silence we often act as if they still exist. As long as issues of race, racial difference, or differing cultural perspectives are viewed as topics to be left unspoken rather than debated, we cannot build a new vocabulary to talk about them. Even those who possess some of the vocabulary cannot continue to build it without honest dialogue. Only through facing the difficulties of the subject can we learn to talk about race or to arrive at new ways of thinking, feeling, and acting. Now, all of us need to contribute in ways that build this lexicon with honest exchanges between and among groups, particularly in families with interracial ties, like mine. To build it, we must talk openly. And we must abandon the myths that cloud our thinking and keep us from the truth, even when that truth is something that makes us uncomfortable.

They say you don't have to choose. But the thing is, you do. Because there are consequences if you don't.

—DANZY SENNA, *CAUCASIA*

II.

The Next Generation

SOMETIMES AS I DRIVE DOWN THE ROAD TO MY GRAND-parents' house, I imagine what it would have been like for a stranger to drive down that same road around 1930, someone like the census taker who probably assumed my grandmother was the cook, not the wife of Jim Richardson. As crafty as Jim was, I'm certain he did nothing to engage with this person or correct the misperception. Instead, he probably signaled in a direct way that the census taker should make his way back down that road faster than he drove up. Although I will never know the true nature of this encounter, I do know that Jim Richardson despised intruders and curiosity seekers. For the sake of the family's survival, the entirety of the Richardsons' life could be open to some but not to all, especially those to whom it might matter how they lived their life and who consequently might bring them harm. Jim Richardson felt that his family life was personal and should not be entered unwelcomingly by anyone outside of his private realm. How the census taker wanted to identify his family really didn't make too much difference to Jim. The fullness of the family's identity was personal, not public.

Racial identity remains a deeply personal issue, yet historically Americans have placed the concept of race on a public, and often political, stage. One of the main platforms on which the concept of race and racial discourse in this country has taken place is the U.S. Census. Throughout American history, the changing nature of race in America has been reflected in the census and the way it collects data on race. In 1790, the first census divided the population on the basis of slave and free, yet race still played a role in the count: free white males, free white females, all other free people, and slaves. When the 1850 census came along, the racial categories had become white, black, mulatto, Chinese, and Indian and remained that way until the "mulatto" category disappeared in 1920, with the rise of Jim Crow. By the 1970s, in the wake of the civil rights movement, the federal government was collecting racial and ethnic statistics on fifteen different racial classifications to enforce civil rights laws related to housing, education, and voting rights.

Given the census's long and tangled history with race and racial demographics, it was no surprise when in the 1990s organized groups of mixed-race Americans began to lobby Congress to add a multiracial box to the 2000 census. Civil rights groups soon cried foul, clinging to traditional racial classifications that had shaped the enforcement of civil rights for a generation. Yet multiracial groups wanted the opportunity to acknowledge the complexity of their racial makeup rather than to choose one racial group over another. At the same time, African American groups felt a multiracial box would divide the black community between dark and light; other groups concluded that while the idea of a multiracial group was interesting, it would not help advance civil rights for minorities. In the end, a compromise was reached: rather than a separate multiracial group, multiracial people could "check all boxes that apply."

Boxes were not checked by the Richardsons; the census taker simply eyeballed the entire family and labeled them white. The family challenged the status quo in their own way, but not with government census takers. While the vestiges of the old identity formulas imposed on the Richardsons persist in American culture, contemporary changes to the census reveal they are now being challenged in a different way.

For my three children, Jim and Edna's great-grandchildren, my wife and I check all that apply and identify them as multiracial. Neither of us wanted to put our children in the position of having to embrace the racial identity of one parent over the other. Nor did we want to imply in any way that one identity was superior to the other. For our children, the idea of a multiracial identity has come to define who we are as a family. While race and racial identity has become tightly woven into the cultural and political fabric of the country, my children are part of a growing population that is redefining what it means to be a person of mixed race. They are part of a movement away from firm racial definitions toward broader racial or ethnic ambiguity.

For the 2000 census, which was the first to allow individuals to "check all that apply" in the category of race, more than 2.4 percent of the population, or 7 million people, reported more than one race. A 2004 Census Bureau estimate shows that 46 percent of this multiracial population is under the age of eighteen, and there are indications that multiracial individuals will make up an even larger part of the population over the coming years. Yet, ignorance and indifference about issues of race and multiracial people continue to divide Americans and perpetuate the demons of our race-obsessed past and jeopardize our moving into a post-racial future.

But as a college student once pointed out to me, "post-racial" holds the connotation that we as a nation have had an introspective discussion and acknowledged the ugly history of race, allowing us to move in a new direction. As a culture we have yet to have this collective new or fresh insight about race, but elements of this much-needed debate are taking place in one of the crucibles of issues of race and identity in this country: American college and university campuses. Most colleges and universities have a myriad number of groups and clubs for every imaginable race and ethnicity. Over the past ten years, students who do not identify with one particular racial or ethnic group but consider themselves multiracial have been organizing their own groups that work alongside other ethnic organizations. "We wanted multiracial people to have a place where they belong and can talk about their issues, which sometimes are not taken seriously," said Phillip Handy, the leader of one such group called Fusion at Rutgers University.

Almost all of the members of Fusion at Rutgers have identified as multiracial all of their lives. "My parents did not encourage me to choose one race. Plus, race is not a big deal to my peers or my family, but since coming to college I have become more confident in my multiracial identity," remarked Dana Sacks. Her words echoed the sentiments of other group members, who all grew up with parents who identified as being of different races or ethnicities but allowed their children to be grounded in both cultures. "My parents never one-dropped us," remarked Sacks, referring to the one-drop rule historically used to define a black identity.

Yet some members of the group find that the perceptions of others place them into specific racial categories, even though they do not see things that way. "Most people understand that I am multiracial, but when I am with black friends I am perceived as black. People make a decision, place an identity on you whether it is your identity or not," noted Fusion member Matt Vaiden. Although they are of different hues and ethnic backgrounds, not simply black and white, they all share common experiences: encountering people who immediately want to determine which racial or ethnic category to fit them into; having elements of their personality written off to racial stereotypes; being aware of the assumption that one parent is not part of the family because of his or her skin color, just as the census taker did with my grandmother; or having difficulty filling out surveys or forms for school. "We might not all be experts on race, but we all see how absurd race can be. From a personal point of view, all of us in this group know how socially constructed race is," remarked Phillip Handy.

With the increase in those who self-identify as multiracial, and with groups like Fusion organizing on college and university campuses across the country, the inevitable question is, can race any longer be a valid way of classifying segments of American society? Checking all the racial or ethnic groups that apply on a census form may provide a more accurate picture of an individual's ancestry, but the issue still arises as to whether having this option is only reinforcing the American obsession with racial categories. While it may be a step forward for biracial and multiracial individuals, is this current reinterpretation of race only a modern manifestation of the one-drop rule, or merely a new mathematical formula for calculating identity, much like the old mulatto-quadroon-octoroon paradigm? Scientists, with the exception of those on the fringes, agree that what Americans define as race is largely a social and cultural construct; yet Americans cling to the concept of race both socially and politically, which seems to perpetuate the myth that race is a meaningful, valid way of classifying individuals. Being able to self-identify as multiracial or multiethnic may be a step forward. But does embracing this racial ambiguity contribute to our society's movement toward a post-racial future or only add a new group to the country's existing racial fragmentation? Does it make this generation more detached from the historic struggles of racism and inequality? Ross Vaiden, Matt's brother and also a member of Fusion, thinks the emergence of multiracial families will help change the country's racial mindset. "The growing number of multiracial families is one of those things no one can control that has to be accepted," he says.

While Fusion and groups of multiracial students are being formed on college campuses, some schools are expanding their definition of students of color to be more inclusive so that multiracial students feel that they belong. In 2003, a group of graduate students at the University of California-Santa Barbara developed a book on multiracial studies to create a broader dialogue about issues faced by multiracial people and to make those issues relevant beyond the multiracial community. In 2007, at Wesleyan University in Middletown, Connecticut, a group called WesConnects started a support network for all students of color, including multiracial students, so that all students could develop an understanding of how their concerns overlap with other groups, rather than seeing them as separated by race and ethnicity. Wesleyan's philosophy has evolved toward recognizing the complexities of racial and ethnic identity rather than reducing it into commonly defined ethnic categories. The former dean for diversity and academic achievement, Daniel Teraguchi, even noted that younger alumni were resistant to being part of alumni groups defined by ethnicity. Like many members of the millennial generation, race is no longer viewed as the only way that people can define themselves.

But if we move away from using race as a valid way of classifying the population, what could be the societal impact of this shift as well as the impact on public policy issues? Racial classification is one way the government measures how racism affects the country in areas such as housing, education, health care, and income, as well as monitoring civil rights violations such as the Voting Rights Act. Moving away from racial classification might mean retooling affirmative action, ending examination of racial balance in schools, and shifting the nature of racial politics. This rising multiracial population might even allow Americans to reclaim some of our melting-pot ideals that have been put into question with the rise of identity politics. All of these ideas make many people uncomfortable, but the questions and debates that an increasing multiracial population brings to our society can no longer be avoided. I faced this first-hand in a message I received from my children's school prior to standardized testing.

"On the test forms, your child will be asked to identify their ethnicity. The choices are: Asian/Pacific Islander, Black (non-Hispanic), Hispanic, American Indian/Alaska Native, White (non-Hispanic). Please make sure that your child knows which option to choose." There was no "check all that apply" category. In a matter-of-fact message to the school, I noted that if "check all that apply" was not an option, my children would not check any boxes. Soon I discovered that I was not the only parent who had protested. As is true in most urban areas, there were a number of other children who also identified as multiracial and were confronting the same dilemma. Based on the number of e-mails the school received, roughly 5 percent of the school's population wanted the option to check several boxes when identifying their ethnicity.

As this example reveals, the current use of race on forms, such as the categories devised by the Census Bureau, by schools, and even on job applications, operates under the premise that ethnic purity is prevalent in American society and that we have a set number of ethnicities as part of our population. That premise is far from the truth. America is a society of polyglots, people who have multiple racial and ethnic backgrounds. The "check all that apply" option on census forms, like the one I worked out with my children's school, at least recognizes that there are people who have parents who are of two or more racial or ethnic backgrounds. Yet, at the same time the race and ethnicity categories used by the census over time have been based on a hodgepodge of criteria, including national origin, language, minority status, and physical characteristics. In many ways, the continued focus on race in census and other forms only adds to the country's already fragmented culture of race and ethnicity.

All of these questions and issues led me to the same conclusion that Matt Vaiden, one of the students from Rutgers, pointed out: we are at a tipping point in this country on the issue of race and identity. With the confluence of a growing multiracial population combined with new perspectives on race gleaned from science, the way we think about race will be changing rapidly and in unexpected ways over the next decade. Simply continuing social policy that is based on the assumption that everyone is either one race or another will not make these changes go away. Multiracial people and young people, Matt Vaiden pointed out to me, "are doing something different in a way that has not been done before, that I hope will be revolutionary and make us see race in a different way." And that is why I believe our social, cultural, and political dialogue on the subject of race needs to evolve to a level that recognizes and acknowledges these revolutionary changes in our society.

My grandparents had to approach the census taker cautiously and accept how he identified their race. Their great-grandchildren can choose to identify as they wish rather than have to conform to how others see them. Without a doubt, the variety of ways in which our culture now frames race and identity will have an impact on society and institutions, such as the census, that use race as a cultural barometer. Perhaps my grandchildren will view the current "check all that apply" convention quaint and not find it necessary to check any boxes at all.

MY THINKING IS OFTEN CAUGHT BETWEEN AN OLD AND A new paradigm: one that acknowledges the absurdity of race and the other that acknowledges its existence as a means of separating people, both currently and historically. In spite of a newly heightened awareness, I can't help but recognize that race played a role in my life. But I like to think that my children's life will be different, their thinking about race less fragmented, not just because of their multiracial background, but because of changing perceptions of race that are moving throughout American culture. Both my father and mother identified as black, in spite of having relatives of every hue and skin color. But it was my mother's family that made a conscious choice to identify as black people, in spite of skin color and features that would have allowed them to move seamlessly into the white world. That choice was socially determined as much as it was personal.

In early twentieth-century Alabama, and across America for that matter, race existed not as something to be transcended as much as it was a marker that determined and defined your social destiny. Racial identity established your station in society: where you could go to school, whether or not you could vote, where you could live, and who you could marry. A combination of law and custom kept racial divisions clearly defined, with no room at all for ambiguity or an existence outside of tightly defined racial boundaries. And if you didn't follow the conventions for choosing your racial identity, there were legal and social consequences. The Richardson family knew what the consequences were and made a choice that embraced a cultural idea whose time was yet to come: that blacks and whites were equals.

The Harlem Renaissance writer Nella Larsen defined passing as a "hazardous business" that required a person to break away "from all that was familiar and friendly to take one's chance in another environment, not entirely strange, perhaps, but certainly not very friendly." What made it unfriendly was accepting a culturally ingrained belief that something deep within your soul, something that was a part of your very being, was inferior. For Jim, taking his family completely into the white world would have signaled to his family and the world that he was superior to Edna. Had he made the choice for the family to pass, he would have succumbed to the very social convention that he despised.

Choosing to be black when you are of mixed race appears to our modern-day eyes as the path of least resistance. Yet the Richardsons took their choice to a different level, making it a radical idea that they actively put into practice in their daily lives. Of course, Jim's power and influence shielded them in ways that still reverberate in the life of the family. In my own life, I know that it was because my grandfather taught my mother to defy Jim Crow that when I was growing up my mother would never let me drink from the water fountain marked "colored." She would always say, "water is colorless, odorless, tasteless," a more sophisticated echoing of what Jim taught her. My mother also taught me that I was equal to whites, even when the segregated world I grew up in told me I was inferior. In social situations I was taught to never defer to a white person. "White does not mean right," my mother often reminded me. Before I realized Jim was white, I thought these lessons came from a proud and strong black man. Now I realize that the lessons she taught me came from a proud and strong man, who in many ways sought to live his life outside of racial categories and boundaries.

I genuinely discovered how much Jim influenced his children's lives, especially my mother's, when at the age of nine my daughter interviewed her grandmother for the StoryCorp oral history project. Before she conducted the interview, I saw it as a once-in-a-lifetime opportunity for Delaney and her brothers, Patrick and Aidan, to understand firsthand that interracial families like ours did not always have an easy time. Our family has had some minor incidents where people stared or pointed at us, but we never faced any danger or direct ridicule. Other people broke those barriers for our family, including my children's great-grandparents. I wanted Delaney and her brothers to get a sense of the passage that had already been cleared for our family from their grandmother, who experienced the barriers set for people of mixed race most directly.

But like most nine-year-olds, Delaney had her own ideas, and the issue of race was not first and foremost what she wanted to know about her grandmother. So, when I sat with her to develop questions, Delaney refused my help. Instead, she insisted on writing her own questions, without either my own or my wife's input. Although I thought the interview would be an exercise in having one of my children understand the way things once were for people of mixed race, Delaney obviously had her own thoughts about what she wanted to know. I deferred to her judgment. In the end, what she wanted to know revealed the differing point of view between generations, both mine and Delaney's and Delaney's and my mother's.

I sat in the background and just listened as grandmother and granddaughter talked. Delaney knew that my grandmother Edna had died young, so she began by asking, "Did your mother ever give you or teach you anything special?" My mother looked directly at Delaney and smiled. "I don't remember that much about her; I was only seven years old when she died," she replied. "She did teach me not to talk back, but my father didn't reinforce it, since he always let me talk back to him."

Then, Delaney asked, "Did your father do anything you were really proud of?" My mother paused for a moment. "He did a lot of things I was proud of." She said this was what she was most proud of: "My father didn't give up on us when my mother died, when at the time he could have. I know that at the time he could have just walked away, but he didn't. And he sent us all to school as long as we would go and paid for it."

Delaney's questions had largely shied away from the issue of race, something I was proud of. Unlike me, she wasn't caught between two ways of thinking. She wanted to get a sense of some of my mother's interests as a young woman, pranks she had played, fights she had had with her siblings, what her everyday life was like, and whether she had ever done anything "really daring" or outlandish. Of course, she had: sneaking into a college dormitory after curfew, and hiding under the blankets of a made-up bed to avoid a trip to Sunday morning chapel were only a few of the things she told Delaney. What my daughter learned in the course of the interview was that Jim had encouraged my mother's sense of daring. "If my mother had lived, my life would have been different," my mother acknowledged. Then, at the end of the interview, Delaney asked a question that I knew was not on her list: "Did people ever make fun of you because your father was white?" Immediately, I felt bad since I felt I had prompted Delaney to ask that question, and I knew it veered from her tightly written script. Delaney had not directly experienced any such ridicule about her own family, yet I learned later that she imagined from books that she had read that this had probably happened to her grandmother. But my mother didn't miss a beat. "If they did make fun of us, they never did it in front of us. If that had happened, I think we would have had a real free for all," my mother said, laughing a bit. "I actually had a pretty sheltered life; Jim really protected us from all the bad things that could have happened to us. We had a very special relationship. That's why, to this day, I can still cry about him. And I know he would have loved to have gotten to talk to you."

Although they had ten more minutes left, Delaney decided to end their talk there. As I sat and watched them talk, I realized how exceptional a life my grandfather had given his children and appreciated how special my mother's upbringing had been. For the Richardsons, living openly as a mixed-race family took a sense of daring, so I was glad that Delaney had learned more about her family's sense of adventure rather than the racial strife they may have endured and often avoided because of Jim. My grandfather largely sheltered them from those struggles; he bore the burden of racial strife personally and did his best to keep it from reaching his children. And in that lovingly sheltered life, he gave his children a sense of well-being, daring, and independence. When the interview ended, I hoped that as a parent I could follow the example set by my grandfather. At that moment, I knew that the real legacy my grandparents left was a vision for helping my children lead a sheltered life where they were allowed to develop a sense of daring.

Memory, in spite of its shortcomings and imperfections, often captures moments in time in ways that satisfy a longing for understanding the past. I believe my mother's memories of her father, while free of many of his shortcomings, objectively capture the essence of the man he was in private. The memories of him that others have shared with me on my travels around Washington County are of the public Jim Richardson, who could be menacing, impatient, and demanding. But he took on that less-than-flattering public persona to make a way in the world that would allow him to provide for his family. "I could get him told and we would still be friends. No one else could tell him off like I could. If they did, they would have been in big trouble," my mother remembered. And that's been the consensus among people I have interviewed: if you crossed Jim Richardson, you had trouble on your hands. Yet, beneath that tough exterior Jim Richardson had a soft heart, especially for a little girl who lost her mother when she was seven years old.

What Delaney's interview with her grandmother exposed to me is the realization of how much race lies on the periphery of the life of my children. Although my wife and I have always openly discussed issues of race, discussions of race have often gone hand in hand with discussions of social justice. Still, race does not play as much of a role in the lives of our children as it did in ours when we grew up in the 1950s, 60s, and 70s. It's a big change from the morning in 1988 after my wife and I became engaged and a restaurant on the eastern shore of Maryland attempted to refuse us service. The message was clear: a black man and a white woman should not be together. The white waitress's pursed lips could barely disguise her disdain. The black kitchen help shot stern "you should know better" stares with their folded arms. A deafening silence hovered over the room.

Since that morning, I know we have been stared at and sometimes have not been treated well because we are an interracial family. Yet my children have no searing recollections of being mistreated. I guess we have our own version of the house at the end of the road, a community where we feel safe, secure, and can be ourselves without fear of interference. Perhaps as a result, so far they don't feel torn between their father's black identity and their mother's Swedish-Irish-Swiss-German background or feel they have to choose one identity or another. Because we have been open about my family's multiracial origins, they have seen their identity as normal. But they all know that there were people who paved the way for them, particularly their grandmother and her parents. And this knowledge has somehow enlightened their worldview in a way that Jim and Edna Richardson could never have imagined.

I think there's just one kind of folks. Folks.

—HARPER LEE, *TO KILL A MOCKINGBIRD*

12.

New Moon Over Alabama

ALTHOUGH PRESTWICK IS NO LONGER THE PLACE IT ONCE was, for my daughter it had not lost its allure. Her grandmother made her think of Prestwick as a place where you could be daring, both outwardly and from within. Though memories of her visit there at the age of eight had dimmed, she still remembered the big white house that appeared at the end of the road like an apparition and the enormous trees that provided shelter from the rising June heat. Three years passed, a score of time for a child, making this tiny spot in Alabama seem distant, exotic, and unknowable, except for what her grandmother shared in their talks and discussions. Somehow I wanted to make Prestwick more real so that she could experience the place afresh, combining her grandmother's memories with what she saw through her own eyes. Although a trip might alter the idealized memory of her last visit, perhaps through what she saw anew we might both look at Prestwick from a new vantage point.

My first trip to Prestwick as an adult was with my children in the summer of 2005. Up until then, I had not ventured into south Alabama since my college days, a time when I felt more like a child than an adult. Although the stories I heard in my teens made my grandparents seem larger than life, my own inner struggles to find my place in the world wouldn't allow me to delve into their lives in any substantive way. Merely piercing the idea of my interracial background felt as if I was compromising my identity as a black man and somehow crossing an invisible line. But by middle age, time had dissipated those insecurities of identity. Staring deep into the mirror, Jim and Edna sometimes seemed to stare back in the curves and lines of my features. And the lessons of my own life made it possible for me to begin thinking of my grandparents as real people rather than as distant figures in my family's history. I wanted these two people to become real, both for me and for my children, since in some ways the pattern of my life and the life of our family lay parallel to theirs.

The path of my life has been far less complex than my grandparents'. I knew that the stares, slights, and nonverbal insults my wife and I have endured as an interracial couple over the years just did not compare with what Jim and Edna experienced. Sometimes it even felt as if they had forged a path for the life my wife and I shared, occasionally guiding us on our way. As our family began to take shape, we cleared an unencumbered path for our own children much as Jim and Edna did for theirs, and the times we lived in made that possible in ways that they could never have imagined. But along the way, something changed: we strived to help our children form a secure personal identity free of the confines of my family's complex and tangled personal history of race and identity. As a result, my own family's narrative seemed to be shifting from one where race stood at its center to one where it held little significance. Personal experience and perspective appear to be shaping my children's identity more than race. Like many of their generation, they see themselves as part of the broader world rather than just their small corner of it, rendering race of marginal significance.

With my own identity secure, by the time I made that first trip back to Prestwick, the real void in my life was my grandmother. I knew that my grandfather's oversized life and legend in many ways concealed her image and sacrifices, sacrifices that allowed him to build that legend. Her absence was carved from the few traces of the life that she left behind with the people who once knew her; there were no pictures or diaries, and the letters she had written had long since been discarded. Talks with my mother cast a mere shadow over the void my grandmother's death had left in her life. In many ways, I internalized my mother's loss and longing. In that process I began to feel a tie to my grandmother that I never thought possible. Almost for that reason alone, I wanted to make my grandmother seem real in a way that would somehow compensate for what my mother lost and what I, in turn, longed to know.

When I came to Prestwick that June afternoon in 2005 with my children, I walked in the house and my mind began to replay the stories my mother told me about my grandparents before our trip began. I thought of the happy times my mother described, especially the games she played with her siblings, as well as the saddest times, which was when her mother died. As I sat in the kitchen with my cousin, Carolyn, while she cooked, we talked about how this was the same kitchen where Edna and Miss Callie worked every day to feed the logging crews as well as the family. The kitchen no longer had the wood-burning stove my grandmother once cut wood for, but it looked pretty much as my mother would have remembered it. Later I walked out to the window where my mother must have peered in to look at the undertaker preparing her mother's body for burial, and I spent time in the room where she died. And we all took turns sitting in the porch swing Jim had made for my mother, a place she comes to rock and reminisce each time she visits. Through contact with these fragments of the past, I began to reconstruct pieces of my grandparents' lives in my mind and form some of the questions about their years together that I wanted to know more about. But on that day, my children did not have those same family touchstones to spark the connections. I'd told them a few stories, but really knew little myself. I had not yet learned enough about my grandparents' life and times to help make them real people in my children's minds. And now that I have spent so much time there and have pieced together even more of their story, I wanted to give them a share of what I had learned about Jim and Edna.

But getting two teenage boys to visit the far reaches of south Alabama proved next to impossible. My sons Patrick, at sixteen, and Aidan, at fourteen, are near the age I was when I learned that my grandfather was white, not black, a memory that remains fresh. As I did at their age, they are striving to form their own identities, free of family and shadows cast by looming forbearers. A trip to Alabama might seem like an imposition rather than an adventure, so I left both of them free to pursue their own journeys. But my now eleven-year-old daughter Delaney agreed with great enthusiasm to go to Alabama. For her the trip still seemed like an excursion to a place that seemed both unknowable and exotic. So the two of us set off to see what we both might find under the heat of the Alabama sun and the cool of its moon.

As with our last trip to Alabama, this one fell during peach season in Chilton County. After Delaney and I left Birmingham we began looking for the giant peach water tower so that we could get fresh peaches to eat during the drive south to Prestwick. Aside from fresh peaches, Alabama's civil rights history seemed to be as much of an interest as family history, since in many ways the two are linked together across time. If for that reason alone, it seemed appropriate that we visit the Civil Rights Memorial in Montgomery as part of our Alabama pilgrimage. The circular granite memorial is inscribed with the names of those who were killed during the civil rights movement, and Delaney stood for quite some time running her fingers across the names as the water trickled down over the stone of the memorial. "I wanted to touch the names that made history and changed my world," she wrote in her journal after we left. At that moment I realized that what changed her world in many ways shaped my worldview, particularly on issues around race. She was moved by walking around Montgomery and, as she said, "getting to walk where Martin Luther King walked." As we looked at the Dexter Avenue Baptist Church where Martin Luther King led the Montgomery bus boycott and the state capitol where the Selma to Montgomery March ended a little more than forty years before, Delaney was enraptured by the history of it all, while I marveled that all of this change happened in my lifetime. Had none of those events taken place, our family would not be able to have the life we lead today.

Like my grandparents, Delaney and I stayed at a house at the end of the road, although it was not the same house. My grandparents' house now is my cousins' home, a place where I am always welcome, but I respect that it now belongs to them. So I sought out a place that would evoke the rustic, remote sense of where her family came from. It was a rustic hunting cabin, one filled with the heads of wild boar and deer, in what was once the town of Carson, just a few miles down the railroad tracks from Prestwick and near where my grandfather grew up. A huge double yoke for oxen hung across the rafters of the cabin, which I reminded Delaney was just like the one my grandfather would have once used in his logging. The cabin's link to Jim Richardson is that it is within shouting distance of his whiskey still, on land owned by his old friend Dan Powell. A map in the cabin labels the spot "whiskey still ridge." "Everybody knows ol' Dan Powell won't take a drink, so the still will be safe," Jim allegedly said when asked why he didn't place the still on his own property. Inside the cabin, you can even find the metal rings that once held the still together.

At night the silence of the cabin was interrupted only by the sounds of tugboats on the Tombigbee River, which we listened for on the front porch. During the day we heard the roar of logging trucks carrying huge cedar and cypress logs from the nearby swamp, just as Jim Richardson did when he logged these same woods. Together we delved into family history, filling in many of the blanks from Delaney's last visit. We began by having a conversation with a newly discovered source. During my last visit, I had found one white Richardson relative who was willing to talk and might shed some light on family relations between the black and white sides of the family. She was a third cousin, well into her nineties, and Delaney agreed to accompany me on the interview. Before we set off, I tried to explain how there were so many different perspectives on Jim and Edna's life and times, many of them shaped by the repressive racial discrimination that sparked the civil rights movement. This interview was yet another attempt on my part to sift through all the stories to try to find some semblance of the truth. During my many trips to Alabama, conversations moved between those who embraced the changes in the racial dynamic of the South and those who remained indifferent to those changes. "It's difficult to get people to talk honestly about race since it makes them uncomfortable," I explained as she nodded in agreement, her face frowned in puzzlement, revealing that she did not quite understand what I meant.

I knew that my explanation of the purpose of the interview was confusing to her, since the world and perspectives I was taking her into contrasted so sharply with the openness she knew in her own life. Delaney had a hard time comprehending why I had faced so much confusion as I sifted through the various bits and pieces of my grandparents' lives. But she dutifully went along on the interview, notebook in hand, to try to help me make some sense of the fractured lives of her great-grandparents.

On my last few visits, I had begun to feel that there was an unbridgeable cultural chasm between the two sides of the Richardson family. Connections with relatives were made and lost, sometimes leading me to hints of despair. But during the interview with my cousin, I began to feel that there was hope, since this newfound relative disputed much of what some members of the family had told me and wanted me to believe: that Jim had only married Edna to spite his mother, that race had never divided the family, and that relationships between blacks and whites, while close and relatively tolerant, were not governed by the prevailing social customs of the South. My cousin reminded me that whites had to call black people "Aunt" or "Uncle," not Mr. or Mrs. "That's just the way it was," she remarked and also noted that in spite of racially driven social customs, most people in Washington County were tolerant of each other. "Everyone down here is related to everyone else, so you can't make too many differences in how you treat other people." During the course of our conversation, she did not mind if she said some things that may have offended me, and we were able to talk through them and share a laugh about the foibles of the Richardson family. Both of us were amused by how the Richardsons had ended up on the wrong side of both the American Revolution and the Civil War yet had intermarried with the descendant of slaves and Native Americans, as if to make up for those lapses of judgment. All of these contradictions made our family uniquely American. In the end, we both recognized that the two sides of the family are more alike than different, since they were shaped by a shared history.

"Washington County was the last frontier and it was more about who you knew that made a difference in how you lived your life. Your grandfather made it here because of who he knew, which is probably why no one ever tried to prosecute him for interracial marriage," my cousin noted. It's what I had thought all along, based on other interviews and archival material, but up until that moment, no one among the white Richardsons had been that direct with me. Although there was much that my cousin did not know, she certainly filled in some big gaps for me. I was thrilled, but I noticed that Delaney was visibly bored. She had stopped taking notes and had begun to stare off into space.

After we left, Delaney admitted that little of what my cousin and I had talked about made much sense to her, so I decided to try to give her some context. We visited Bassett's Creek, where my grandfather's forbears became the first settlers in this part of what was then the Mississippi Territory. And we went to Saint Stephens, where my grandmother grew up to see the remains of the town and talk with archeologists who are searching for the site where our ancestor, the freewoman of color Mahala Martin, once lived. I explained that Mahala Martin was known to the whites of Saint Stephens in the 1820s as "Aunt Hagar," echoing my cousin's remark that black people were always known as "Aunt" or "Uncle." Because of the significant role Aunt Hagar played in the history of Saint Stephens, archeologists were trying to find signs of what her life was like through items that may have been left behind.

And we visited the cemeteries where both of my grandparents are buried. We visited my grandmother's grave first, and the idea that a married couple could not be buried on the same plot of land made little sense to Delaney. But when we arrived at Pine Grove Cemetery, where my grandfather is buried, and she saw the graves of relatives who fought in the Civil War, it seemed to make sense. The presence of small Confederate flags in the cemetery signaled that this was a place where her grandmother could not have been buried at the time. Delaney remembered Confederate flags in the hands of whites jeering blacks in civil rights protests, so seeing them in the cemetery made it clear why her grandmother was not buried there.

After leaving the cemetery, we headed to my grandparents' house to sit on the porch swing for a while and to talk about some of the places we had been as well as what life must have been like at the house when they lived there. I explained that the house and land were deeded to Edna from Jim, since they could not own the property together because of laws against interracial marriage. And I explained that Jim built the house specifically for Edna. "Then he must have really loved her," Delaney said. I agreed and told her how unusual it was for a woman to own property then, much less a black woman. But Delaney was confused. "Your grandfather built this house for your grandmother. Obviously they loved each other. So why do some people say he married her to spite his mother?"

"I've been coming down here the last few years just trying to figure out your great-grandparents' lives and to find out who they really were. I've learned a lot about them, but I've also learned that sometimes people choose what to believe, even something that is not true. And some people have a hard time believing that two people of different races could actually love each other." The answer seemed to make sense to her. Still, I knew that this talk of race was perplexing and foreign. No one she knows feels compelled to think of people in terms of race. She told me that she got bored during the interview because to her it didn't matter who was black or white, and she couldn't really understand why it was so important.

On the way back down the road, Delaney asked me to stop, since she wanted to see up close how the moss hung from the trees that shaded the road. The heavy gray strands seemed so exotic, like nothing she had ever seen before. At that moment, I remembered stopping near the same spot with my mother when I was a small boy, as she pulled moss from the trees to hang on the trees at our house in Mississippi. My mother must have needed to have a reminder of home near her, albeit briefly, since the moss never took to the mimosa trees on our farm. Watching Delaney, I realized that I've never seen another road that looked quite like this anywhere else in the world. "Prestwick was a unique place," my mother once told Delaney. Gazing down the road, I finally understood what she meant.

As Delaney pulled the moss from the trees, I remembered that this spot is one of the few parts of the road still completely shadowed by moss. It was quiet, as I am sure it always has been, even when there was a lively community nearby. In the quiet, as I snapped photos of Delaney, I reminded myself that I came back to Prestwick to be a custodian of memory, particularly memories of this place. If all Delaney remembers from this trip is gathering moss, just as I remember my mother doing the same thing, she is a custodian of that memory along with me, which is something we will always share.

We spent only one more day around Prestwick, since Delaney wanted to get back to Birmingham to see the Civil Rights Institute and the Sixteenth Street Baptist Church, the scene of many of the civil rights mass meetings of the 1960s. I wanted to stay longer, but sanguinely knew that our time in Washington County had to come to an end. As we left, I realized how barren Prestwick must look to Delaney. Over the years I've had the opportunity to fill in the blanks in my mind and can imagine how vibrant it once was: logging crews, my grandfather's runs to the whiskey still, children running down the moss-covered road to school, Sunday gatherings at the train depot, services at the now-abandoned church. That's all gone now, and there are no children here to bring it back, only people whose lives and memories are fading. Still, all those stories bounce against each other in my mind, and I have often placed that tapestry of memories on top of the remnants of what is left of Prestwick.

The drive back to Birmingham took us through Perry County, which Walker Evans photographed during the Great Depression for the Farm Security Administration. Unfortunately, no one documented Prestwick in photographs as they did other small towns around here. The landscape of Perry County has changed little from the 1930s and seeing it helped me keep my time in Washington County locked in my memory.

Though I wanted to visit the Civil Rights Institute, I longed to find the Sprott, Alabama, post office, the setting of one of my favorite Walker Evans photographs. The Sprott post office bears a striking resemblance to the old general store and depot in Prestwick, as I remembered seeing it as a child in its somewhat decrepit glory. Going to Prestwick always seemed like going back in time, back to the poignant simplicity of the Depression era my parents sometimes talked about. But now much of what evoked that sense of the past is gone. Seeing Sprott would have stimulated my memories of old Prestwick, but I resisted. Old Prestwick is gone and undocumented, and nothing can bring it back.

It wasn't until the curtain went up on the orientation film at the Civil Rights Institute, taking us deep into the world of Birmingham at the height of the Jim Crow era, that a flash of truth radiated back at me from the museum objects that came into view. The colored water fountains and Jim Crow signs immediately reminded me of my childhood during segregation. And those signs were exactly the same ones my mother and grandparents knew as well, many of them put into place in the 1920s. While I have been trying to get Delaney to think about the role race played in my life and the generation before me, she had difficulty comprehending all the historical events and changes we talked about in Prestwick because the conversations took place from a Depression-era perspective of race. Delaney, her two brothers, and the rest of their generation look at the events of the past through the lens of the civil rights movement, the era that wiped those signs away. Jim Crow signs for them are history, not memory. All of the events of the last generation, combined with living a life free of the old confines of race, led them to develop a perspective shaped by a sense of social justice rather than race. While I have been trying to comprehend my family's history and get to the point where I can embrace the common humanity of my relatives, rather than to distinguish them by race, Delaney and many of her generation are already there. And it was the events of the civil rights movement that has gotten them there. For Delaney and her generation, race relations before the civil rights movement, particularly during the Depression, seem too distant to be knowable in any real way, except perhaps through objects in a museum.

My children's generation contrasts sharply with mine. My generation was the one that embraced blackness and changed it from a source of shame to one of pride. A black identity was central to whom I was, and I wanted to project that identity in every possible way: hair, dress, and political beliefs. By contrast, Delaney and her generation acknowledge the existence of race and ethnic identity, yet it doesn't rest at the center of their being. What matters more is justice, and justice matters because we are all human and should not singled out for things that really don't matter, like race. She's proudly multiracial and wants to acknowledge every facet of her background and won't confine her identity to one group or another. It's just not something she or her brothers feel a need to do.

I took Delaney to Alabama to teach her about the multi-racial origins of her family. Though I would like to think that memories of our trip to Prestwick will stay with her for the rest of her life, Delaney taught me more than I was able to teach her. She taught me that in some ways I'm still far too locked into my generation's perspective on race and racial identity, which in many ways was shaped by the generation before mine. In my generation, everything was seen in terms of black and white, to the exclusion of any other perspective. Delaney and many of her generation are rejecting that view. If I want to move into the future alongside Delaney and the rest of her generation, I need to truly transcend race. That means I need not embrace my identity so tightly that it locks out any other perspective. I'll always be there to remind her of the lessons of history, a history that I lived and that was steeped in the culture and politics of race. But what she has shared with me is a vision of something I never could have imagined at her age: a world where race matters less and justice and our common humanity matter more. And I truly believe that her generation is going to help us get there.

The importance of justice and equality, not race, is the real lesson of the life of the Richardson family. When I made that first trip to Prestwick as an adult, I thought that in time I would take away a detailed portrait of Jim and Edna, painted from the fragments of memory. Instead, I have a clearer sense of what their lives represent, both to me and my children, and how their past bridges to my present and future, a future that moves beyond race. Rather than a portrait, I have something more meaningful that allows me to witness their life each day vividly, in full view, and from a fresh perspective, free of many of the burdens they carried to their house at the end of the road. And I think that is what they would have wanted.

Epilogue

EACH TIME I LOOK AT THE PORTRAIT OF JIM RICHARDSON that hangs on my wall, I long to see a picture of Edna beside him. Since none exists, over the years I've tried to bring one into focus in my mind's eye. In my wanderings around Prestwick, numerous people have described Edna in vivid language, detailed enough for a gifted painter to create a warmly inviting image. But with my limited visual skills it's been difficult to bring her into focus on the canvas in the deep recesses of my mind. The only way I have found to imagine her is to look at Jim's portrait and an old photograph I have of my mother as a very young woman. In the same long oval space as Jim's portrait, the composite my mind melds together from these two images is still blurred. Nevertheless, the image is captivating enough to satisfy a long pent-up need and longing.

When you set off in search of missing pieces of your personal history, what you find is not at all what you set out to discover. While there is much that I've learned about Jim and Edna, there is much that I can never know and can only infer from the times in which they lived. Over the years I've learned some of their sorrows and a few of their joys, though none of them completely. Instead, what I have discovered is how the choices and decisions these two people made in the not-so-distant past, in a place that no longer exists, affects the way I live now. In my quest to find out more about Jim and Edna, in their fractured and brief lives, I learned lessons for my own life and family today.

What I learned is that while race determined the course of their lives, at the same time they broke free of its burdens as best they could. While they fought and struggled with each other, they formed a family structure that allowed their children to develop an identity that was true to themselves, but would also guide them through the wider world. Identity meant who you felt you were inside, and it was something of which you were never to be ashamed. Within that identity there was no room for self-hatred, doubt, or superiority. Everyone was your equal, there to be judged on individual merit, not by how white or black they were. Equality was the path to achieving justice.

In many ways, these two people, by all accounts both tough and combative, carved out a path for my family to exist in full. Yet I know I cannot pass all their lessons in identity down to my children. Times have changed and there is much more freedom in this century to develop a personal identity that can begin to move beyond race. That's not exactly the lesson they left hidden for me as I tracked down the broken pieces of their lives, but it is what I have learned from them and is what I have chosen to pass on.

In the pieces of Jim and Edna's life, I've found a new story for my family to take forth into the ages. It is one that places us and our descendants on the precipice of freeing ourselves from some of the burdens of race, particularly those that separate us from others just because they look different or hold a belief system counter to ours. I know we won't get there completely; our cultural identities are what make us feel secure and safe. I know that's why I held so tightly to my identity as a black man when I learned that one of my closest relatives, and the one dearest to my mother, was a white man. That's why I continue to hold on to that identity. But along with my family I plan to try in good faith to move beyond race because by doing that we are keeping a family legacy alive in a way that means more than a portrait I can hang on my wall. It's a way to keep shining whatever light Jim and Edna brought to this world.

Along the roads of Prestwick lie the vestiges of the light Jim and Edna brought to this place. You hear it in the voices of the people who once knew them. It breaks through in the cracks of their rapidly aging voices as they tell the same stories over and over again, almost the same way the sun shines through the gray moss on sunny days and reflects in the shimmer of the sandy road. Sometimes as I walk along the road, I stop and stand in places I imagined they stood or walked. While I turn my thoughts over to the quiet that surrounds me, if I listen carefully I will hear Edna calling to Jim from the creaky front porch. Glancing over at the house, I see a little girl, my mother, looking at Edna admiringly and lovingly from the porch swing Jim made just for her. Edna's voice is sharp, pointed, direct, and no nonsense, as is Jim's in reply. As the light fades, in the chords of this imagined call and response I can still feel their love for each other as I look at the house that stands as a memorial to both of them.

And that is my vision. It can never take its place on the wall beside Jim's portrait. But it is full and complete, much like the lives Jim and Edna lived.

THE EUBANKS FAMILY TODAY

Selected Reading

Writing about race and identity, even from a personal perspective, would have been impossible for me without reading much of the work of other writers on this topic. I owe a special debt of gratitude to the scholarship of Kwame Anthony Appiah, whose writings inspired me to think about race differently as well as to begin to look beyond race in my everyday life.

Appiah, Kwame Anthony (2006). *Cosmopolitanism: Ethics in a World of Strangers*. New York: W.W. Norton.

Appiah, Kwame Anthony (2005). *The Ethics of Identity*. Princeton, NJ: Princeton University Press.

Appiah, Kwame Anthony (1992). "Illusions of Race," in *In My Father's House: Africa in the Philosophy of Culture*. New York: Oxford University Press.

Blight, David W. (2001). *Race and Reunion: The Civil War in American Memory*. Cambridge, MA: Harvard University Press.

DaCosta, Kimberly McClain (2007). *Making Multiracials: State,*

Family, and Market in the Redrawing of the Color Line. Stanford, CA: Stanford University Press.

Didion, Joan (1979). *The White Album*. New York: Simon and Schuster.

Ellison, Ralph (1952). *Invisible Man*. New York: Random House.

Hill, Lawrence (2001). *Black Berry, Sweet Juice: On Being Black and White in Canada*. Toronto: HarperCollins.

Kennedy, Randall (2003). *Interracial Intimacies: Sex, Marriage, Identity, and Adoption*. New York: Pantheon.

Malcomson, Scott L. (2000). *One Drop of Blood: The American Misadventure of Race*. New York: Farrar Straus and Giroux.

Novkov, Julie (2002). "Racial Constructions: The Legal Regulation of Miscegenation in Alabama, 1890–1934," *Law and History Review*, Summer 2002, http://www.historycooperative.org/journals/lhr/20.2/novkov.html (27 Feb. 2008).

Novkov, Julie (2008). *Racial Union: Law, Intimacy, and the White State of Alabama, 1865–1954*. Ann Arbor: University of Michigan Press.

Scharfstein, Daniel J. (2003). "The Secret History of Race in the United States." *Yale Law Journal*, Vol. 112, No. 6 (April), pp. 1473–1509.

Sen, Amartya (2006). *Identity and Violence: The Illusion of Destiny*. New York: W. W. Norton.

Senna, Danzy (1998). *Caucasia*. New York: Riverhead Books.

Severo, Richard (1970). "The Lost Tribe of Alabama," *Scanlan's Monthly*, pp. 81–88.

Sims, George E. (1989). *The Little Man's Big Friend: James E. Folsom in Alabama Politics, 1946–1958*. Tuscaloosa: University of Alabama Press.

Sokol, Jason (2006). *There Goes My Everything: White Southerners in the Age of Civil Rights, 1945–1975*. New York: Alfred A. Knopf.

Sollors, Werner, editor (2004). *An Anthology of Interracial Literature: Black-White Contacts in the Old World and the New*. New York: New York University Press.

Sollors, Werner, editor (2000). *Interracialism: Black-White Intermarriage in American History, Literature, and Law*. New York: Oxford University Press.

Spencer, Jon Michael (1997). *The New Colored People: The Mixed-Race Movement in America*. New York: New York University Press.

Spickard, Paul R. (1989). *Mixed Blood: Intermarriage and Ethnic Identity in Twentieth-Century America*. Madison: University of Wisconsin Press.

Tate, Allen (1938). *The Fathers.* Athens, OH: Swallow Press/ Ohio University Press.

Taylor, Charles (1989). *Sources of the Self: The Making of Modern Identity.* Cambridge, MA: Harvard University Press.

Torrence, Ridgely (1948). *The Story of John Hope.* New York: Macmillan.

Acknowledgments

When I thought about writing this book several years ago, I never could have imagined the path it would lead me on, both as a writer and a human being. Casual conversations by the side of the road or on a front porch evolved into interviews and, over time, friendships. As I sat in solitude to write and contemplate the conversations all of these new friends and relatives shared with me, I never felt alone. Words spoken and unspoken became a part of me and influenced my writing as well as my thinking. I am extraordinarily grateful to all who took the time to travel along with me.

The uninterrupted time to write this book would not have happened without the freedom provided by a fellowship from the John Simon Guggenheim Memorial Foundation. A special thanks to the staff of the foundation and to Ann Abadie, James Ronald Bartlett, William Ferris, and David Levering Lewis for their support in pursuing the fellowship.

The New America Foundation gave me a home when I needed additional time and support to write. I am especially grateful to Steve Coll and Sherle Schwenninger, who saw not only a personal story but also one that has an impact on modern American culture and social policy. My newfound colleagues at New America also provided much support, feedback, and encouragement and include Andres Martinez, Michael Lind, Ray Boshara, Mark Schmitt, and Philip Longman.

Other institutions gave me the opportunity for quiet reflection as well as a forum to read from portions of *The House at the End of the Road* as it progressed. I am grateful to Adam Goodheart and the staff of the C. V. Starr Center for the Study of the American Experience at Washington College in Chestertown, Maryland, where I served as the 2007 Frederick Douglass Visiting Fellow. The Starr Center gave me a quiet place to write as well as a forum to interact with potential readers. The Alabama Writers Symposium in Monroeville and Millsaps College in Jackson, Mississippi, not only gave me an opportunity to read from my work in progress, but also allowed me a chance to me to lecture about my writing and research. I am especially grateful to Suzanne Marrs and the staff of the Eudora Welty Foundation for making my week at Millsaps so memorable. The thrill of standing behind Eudora Welty's writing desk will never leave me as long as I live and helped propel my work on several chapters of this book.

This journey back to my grandparents' house at the end of the road would not have been possible without the loving support and sacrifice of my family. As I wrote, they all served as the gentle reader over my shoulder, a role of great importance. My wife, Colleen Delaney Eubanks, always listened patiently (even when I did not), helped keep my research organized, and assisted in my balancing of writing and the routines of domestic life. My sons, Patrick and Aidan, dutifully read early drafts and provided comments and insights that revealed maturity, intellectual depth, and wisdom beyond their years. I owe a special debt to my daughter Delaney's interviews, questions, and probing curiosity. Each day all of you teach me something new.

Although I could not have conversations with Jim and Edna Richardson, my mother Lucille Richardson Eubanks and her surviving siblings, Mary Richardson Edwards and R. J. "Smokey" Richardson, served as their surrogates. All of your stories and insights made Jim and Edna's world come alive for me and unlocked a world I knew little about when I started. Thank you all for your honesty and the tears we shared when it pained you to talk.

One of the joys of writing this book was getting to spend time with cousins, both old and newly discovered. Drives around Washington County were fueled in part by the fresh pound cake Carolyn Jenkins always seemed to have waiting for me each time I visited. Without Jimmy Jenkins, I would have been lost on many a country road. Thanks, Jimmy, for all of your help navigating both roads and the people we encountered together. Your gentle grace and patience was an inspiration to me. Thank you Laura Cooper and Janat Clemendor for sharing your Prestwick network with me and connecting me with all the right people. Thanks also go out to Bill King, Joyce Hogi, and Chlora May King, who helped put me in touch with the few people left who remember the glory days of Prestwick.

So many members of the Richardson family opened their doors and hearts to me: Kay Collier, Reuben Powell, Billy Richardson, Roy Richardson, Woody Richardson, and Margaret Richardson Thames. Thank you for your graciousness. We are all family now. And a special thanks to Pat Foster, who shared not only her stories, but her innermost thoughts and unwavering support.

Numerous people in and around Washington and Clarke counties sought to lend a hand in some way, providing warmth and support that truly made me feel at home: Harvey and Elizabeth Jackson opened their home and social network to me, connecting me with Lee Jackson, who shared stories of helping Jim Richardson near the end of his life on his legal affairs; I also received help and support from Tammie Chapman, Mitt Connerly, Joseph Franklin, Vivian Gregg, Darlene McIlwain and the staff of the Creek Bank Restaurant (who kept me supplied with the best cheeseburgers in ALL of Alabama), Nap Roberts, Judge Tom Turner, L. J. Williams, Wyoline Woodyard. And in memory of Shed Gregg and Tom Pruitt, who generously gave me their time during several interviews but did not live to see this book come to be. It was truly an act of God's grace that I came to know both of these men.

Roy Hoffman's profile "A Grandson's Quest" in the Mobile *Press Register* gave me a wealth of sources to tell this story. Thank you for your support and friendship.

A special thanks to Will Powell as well as Billy and Matt Powell, whose hunting cabin in the wilds of Carson, Alabama, became my writing refuge. The rustic setting and Bo Walker's dinners of grilled wild boar made the world of frontier Washington County real rather than just pulled from my imagination. Bernice Bolling Williams gave me a fresh view of Prestwick with her exceptional visual memories of the place and its people.

Archivists and historians became close associates as I pulled together the pieces to this story. Jim Long of the Saint Stephens Historical Commission seemed to always step in with a source or a historical insight that kept my work moving along when I felt discouraged. Wanda Braun, also of the Saint Stephens Historical Commission, always stood by to help when Jim was not available. Both her historical knowledge of Washington County as well as her blood tie to the Richardson family proved to be an invaluable resource. At the University of South Alabama Archives, my thanks go to Carol Ellis and her team of archivists, who took a special interest in my work. At the Alabama Department of Archives and History, special thanks to Norwood A. Kerr for his help with my research on miscegenation cases. Julie Novkov provided a broad window into the Alabama's miscegenation laws, and Wayne Flynt provided guidance on Alabama history. I'm grateful to Phillip Handy, Dana Sacks, and all the members of Fusion at Rutgers University, who took time out of their busy schedules in the middle of exams to talk with me, and to Nicholas Pastan for introducing me to his generation's perspective on race. Hardy

Jackson's comments on my first draft proved to be invaluable, and I am grateful for his generously sharing his knowledge and time with me.

As always, I owe a special debt of gratitude to my agent and friend, Martha Kaplan, who always believes in my work even when I do not. My dear friend Frederick Reuss gave me gentle but pointed comments on my early proposal and helped me find my subject. My editor, Elisabeth Kallick Dyssegaard, shaped the manuscript and took me into the world of my potential reader when I was often on another plane. Writers cannot survive without good editors, and you are one of the best. As always, Evie Righter's artful copy editing smoothed out the rough spots.

Finally, a special note of thanks to Jim and Edna Richardson. Thanks for always being there to walk at my side along the roads of Prestwick.